D0895309

Seismic Stratigraphy

Seismic Stratigraphy

Robert E. Sheriff

INTERNATIONAL HUMAN RESOURCES DEVELOPMENT CORPORATION

Boston

 For information address: IHRDC, Publishers, 137 Newbury Street, Boston, MA 02116.

ISBN: 0-934634-08-4

Library of Congress Catalog Card Number: 80-83974

Printed in the United States of America

Contents

Preface

Every little wiggle has a meaning all its own. This is our underlying faith, that details of seismic waveshapes can tell us the details of the nature of the earth. But their voices are obscured by many irrelevancies. They speak in a high-noise environment, and we have been able to decipher only a small portion. However, things are looking up: better techniques are lessening the irrelevancies, and we are learning to read.

In exploration of unknown areas, determining the nature of the rocks present is often the difficult aspect. Most of the properties of rocks that can be measured at a distance are not distinctive enough to identify the rock unambiguously. Conventionally, seismic data are used to determine aspects of the structure. Stratigraphic pictures are inferred from the structure, the nature of rocks exposed for examination in the surrounding area, and regional concepts.

Three points make seismic stratigraphy feasible now: (1) we have better data quality, (2) we have begun to systematize analysis procedures, and (3) we believe in the geologic significance of waveshape details.

This book is an overview of the analysis procedures of seismic stratigraphy. It was written as a companion to a set of videotapes. The content is essentially that of the tapes, but it does not bear a one-to-one correspondence. The viewer is discouraged from reading the book while viewing, since the book is not a transcript of the tapes. The book presents some viewpoints in greater detail, in other instances with less discourse. Sometimes the same material is approached in a different sequence or with a different line of reasoning. Some illustrations on the tape are not included and some illustrations in the book are not on the tape.

Acknowledgments

Seismic stratigraphy owes much to a greater awareness on the part of interpreters that seismic data contains useful stratigraphic information, an awareness partially attributable to the AAPG-SEG Schools on Seismic Stratigraphy. These schools systematized procedures for use in seismic stratigraphic analysis. The AAPG-SEG schools evolved from a research symposium on seismic stratigraphy held at the 1975 AAPG convention in Dallas. AAPG Memoir 26, *Seismic Stratigraphy—Applications to Hydrocarbon Exploration* (C.E. Payton, ed., 1977), developed from the Dallas meeting and the schools; today this book constitutes the principle literature on the subject. Most of the illustrations in the present work derive from this work.

Acknowledgment is made to the American Association of Petroleum Geologists and especially to P.R. Vail, R.M. Mitchum, R.G. Todd, J.M. Widmier, S. Thompson, J.B. Sangree, J.N. Bubb, and W.G. Hatfield—the Exxon contributors to AAPG Memoir 26. Acknowledgment is also made to L.F. Brown, W.L. Fisher, M.B. Dobrin, M.T. Taner, N.S. Neidell, L.D. Meckel, J.P. Lindsey, W. Laing, R.O. Lindseth, and others who have been used as sources for this work and who have made important contributions to the AAPG schools, and to companies such as Exxon, Conoco, Geoquest, and Seiscom Delta and others for their support of the speakers in these schools. Appreciation is also expressed to Judy Golasinski and her colleagues in the AAPG education department; the smooth operation of the AAPG schools have not only benefitted the attendees but also the speakers by providing a forum where they could periodically interact and learn from each other. The contribution of figures and data by companies who elect to be anonymous is also acknowledged.

Dedication

To Milton B. Dobrin—Dr. Dobrin was one of the original teachers in the AAPG Seismic Stratigraphy Schools, and he instituted at the University of Houston what may have been the first academic course in seismic stratigraphy. He was a great scientific investigator, teacher, humanitarian, and friend.

Seismic Stratigraphy

PRINCIPLES

GEOLOGY:

CHARLES LYELL, Esq. F.R.S.

VOL. I.

THE FIFTH EDITION.

LONDON:

JOHN MURRAY, ALBEMARLE STREET.

Figure 1.1. Frontispiece and title page of Lyell's *Principles of Geology* (fifth edition, 1837). The Temple of Serapis shows the record of several water-level stands of the Nile.

1
Introduction

Sedimentary rocks normally exist in more-or-less parallel layers or strata, different rock types being distinguished by different physical features. Lyell (see figure 1.1), in *Principles of Geology* (1830), quotes Strabo as writing in *Geography* about A.D. 20:

> It is not, because the lands covered by seas were originally at different altitudes, that the waters have risen or subsided or receded from some parts and inundated others. But the reason is, that the same land is sometimes raised up and sometimes depressed, and the sea also is simultaneously raised and depressed, so that it either overflows or returns into its own place again. We must, therefore, ascribe the cause to the ground, either to that ground which is under the sea or to that which becomes flooded by it, but rather to that which lies beneath the sea, for this is more movable . . . [and] can be altered with greater celerity.

Determination of relative sea level is one of the features which seismic stratigraphy seeks to ascertain from seismic data.

Lyell defined *strata*:

> Strata: The term stratum, derived from the Latin verb *struo*, to strew or lay out, means a bed or mass of matter spread out over a certain surface by the action of water or in some cases by wind. The deposition of successive layers of sand and gravel in the bed of a river or in a canal affords a perfect illustration both of the form and origin of stratification. A large portion of the masses constituting the earth's crust are thus stratified, the successive strata of a given rock preserving a general parallelism to each other, but the planes of stratification not being perfectly parallel throughout. . . .

The American Geologic Institute *Glossary of Geology* (1972) defines *stratigraphy* as:

> The branch of geology that deals with the definition and description of major and minor natural divisions of rocks . . . and with the interpretation of their significance in geologic history, specifically the geologic study of the form, arrangement, geographic distribution, chronologic succession, classification, and especially correlation and mutual relationships of rock strata . . . in terms of their origin, occurrence, environment, thickness, lithology, composition, fossil content, age, history, paleogeographic condition, relation to organic evolution, and relation to other geologic concepts.

Stratigraphy thus embraces most of what one might consider as the geology of sedimentary rocks. We wish to deduce as much of this as possible from seismic data.

Figure 1.2 shows the waves on the surface of a pond that result from dropping a pebble into the water. The waves that have been reflected from the rock carry information about the contrast at the water-rock interface. Seismic reflections likewise carry information about the interfaces between rock types in the earth.

Lateral changes in rock type are the general rule, although usually rock type changes occur gradually. For example, sediments deposited near their sources are generally coarser

Figure 1.2. The reflected waves carry information about the configuration and contrast at the water-rock interface.

grained than those deposited farther away, and we expect a formation that is mainly sandstone to gradually change into a shale as we move away from the source. We also expect other types of formations to change their nature along the bedding. Changes in rock type produce changes in the reflectivity, which affect the wave shape seen in seismic data. Inferring stratigraphic changes and where they occur based on characteristics of seismic data is an objective of seismic stratigraphy.

While most of our pictures of rocks come from their exposures on land, seismic stratigraphic analysis is also done with marine data. Data recorded at sea are usually more uniform than land data because the recording environment is more homogeneous and produces fewer distortions, so that it is easier to see variations attibutable to stratigraphic changes. However, the principles of seismic stratigraphy are the same whether the data are recorded on land or at sea, and many of the examples in this book are based on land data.

The high-amplitude event shown in figure 1.3 is produced by a hydrocarbon accumulation that has changed the properties of the host rock sufficiently to affect the amplitude of the reflection from it. "Bright spots" or "direct hydrocarbon indicators," as such evidences are called, are now regarded as a subset of seismic stratigraphy (see chapter 9).

The most obvious information obtainable from seismic data is the configuration of interfaces, and the historical use of seismic data has been in the mapping of geologic structure. Most stratigraphic conclusions from seismic data have been inferences from the structure that the seismic data determine, and this type of reasoning is still extensively used in inferring stratigraphy. The distinction of seismic stratigraphy is that it goes beyond merely mapping structure and searches for more direct seismic evidence as to the nature of the rocks and the fluid contained within their pore spaces. Many of the features in the above definition of stratigraphy

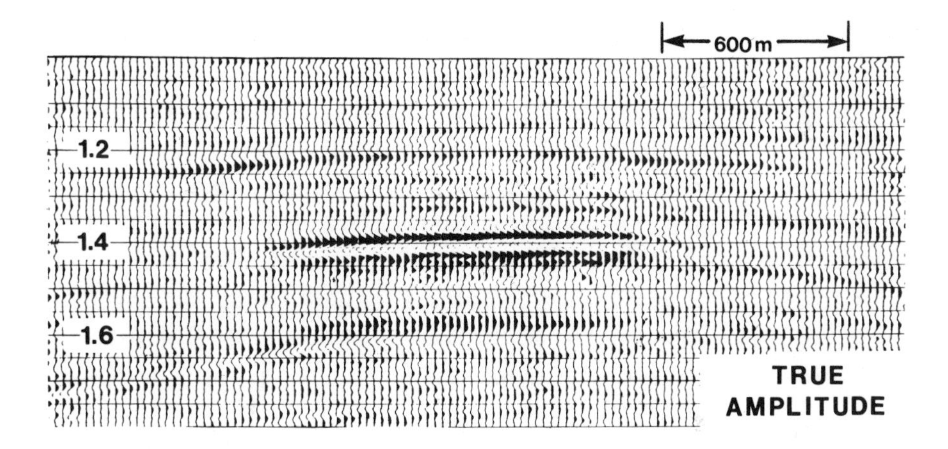

Figure 1.3. High-amplitude event indicating a hydrocarbon accumulation. (From Schramm, Dedman, and Lindsey, 1977; reprinted by permission of The American Association of Petroleum Geologists)

show evidences in seismic data, and tentative conclusions about many stratigraphic features can be drawn from seismic data.

Stratigraphic Patterns in Seismic Data

Among the more obvious stratigraphic features are structural patterns such as those shown in figures 1.4 to 1.13. Figure 1.4 shows generally flat-lying reflections but a belt of dipping events forms a progradational pattern across its center (AA′); this is interpreted as indicating original depositional dip in the outbuilding of a delta. Above this can be seen an unconformity BB′, shown by the transgressive onlap of seismic reflections from the left. Still higher in the section (at about CC′) there is a strong change in reflection character. The reflections above CC′ are higher frequency, have less continuity, and are more "wobbly" than those below; the interpretation is that those above CC′ represent nonmarine deposition, those below represent marine. A present-day sea-floor channel adds some distortion and interruption of continuity; we must be certain that we do not misinterpret the deep effects of shallow features as indicative of stratigraphic changes. The strong reflections and strong continuity of the lower part of the figure represent mainly marine deposition. The structure that produces the reversal in horizon DD′ was apparently positive at the time when the next overlying sediments were deposited, as shown by their thinning onto the structure. Seismic stratigraphy thus seeks to determine paleostructure and paleoenvironment.

Figure 1.5 shows a present-day continental shelf edge. A number of former shelf edges can be seen in the reflections at greater depths. The pattern is so dramatic that it is easy to pick periods of mainly outbuilding, periods of mainly upbuilding, and periods of uplift when the tops of patterns were being eroded.

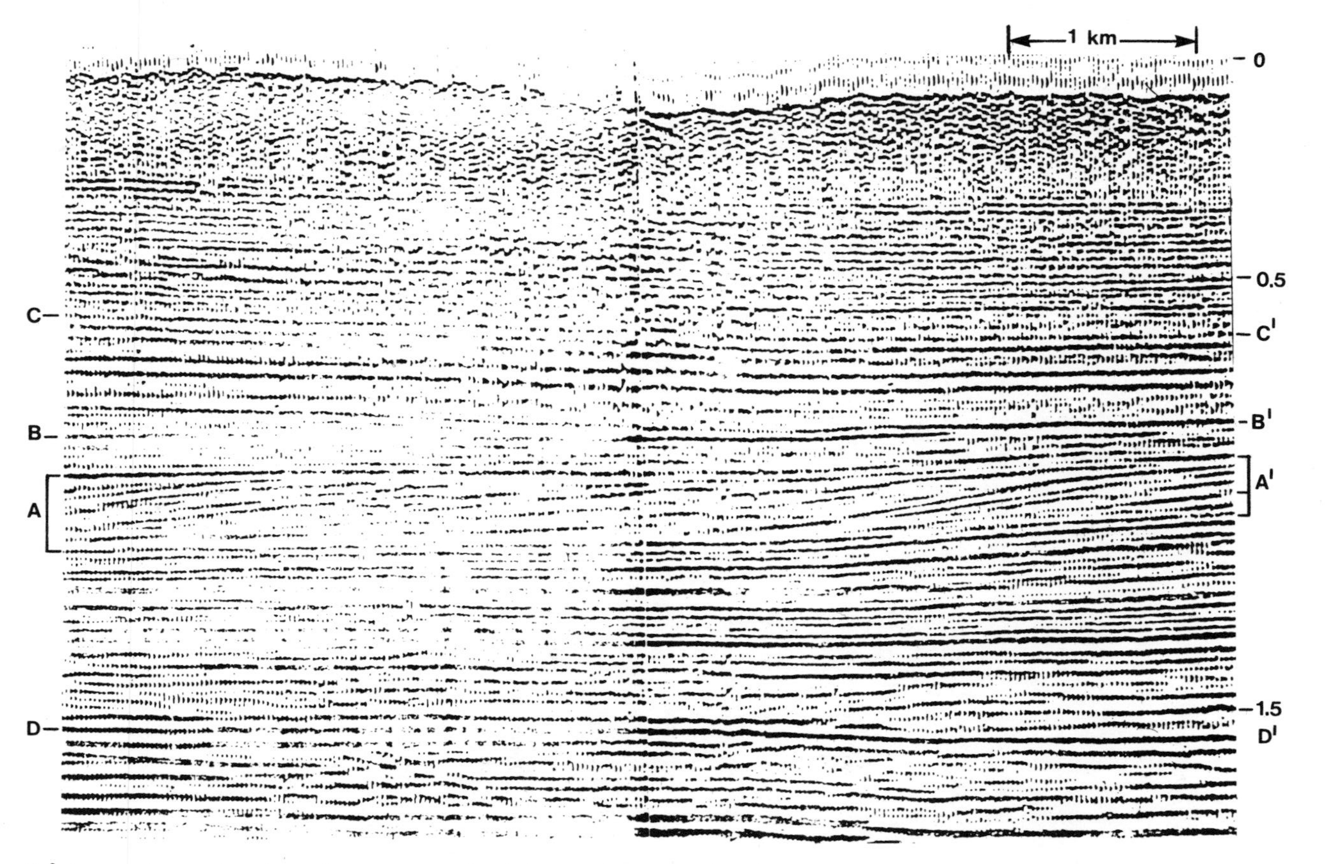

Figure 1.4. Section showing outbuilding of a delta (AA′), unconformities (BB′, DD′, and elsewhere) and other stratigraphic features. (Reprinted by permission of Chevron Oil Company)

Transgression onto an unconformity is the obvious feature of figure 1.6. There is some thinning to the right of the onlapping units, indicating slight subsidence and tilt to the left during their deposition.

Figure 1.7 is a migrated section of a tight fold. Some of the intervals between reflections maintain constant thickness through the fold; these are the competent units. Other units thin and thicken; these are the shales. Observations of the response of rocks to stress gives information about the nature of the rocks.

Another set of fairly obvious patterns is shown in figure 1.8. The overall pattern is of horizontal reflectors but the zone AA′ shows not only appreciable relief on its top but several distinctly different reflection patterns within it. The no-reflection zone in the center is the Horseshoe Atoll reef, with back-reef facies to the right and fore-reef facies to the left.

Figure 1.9 shows several patterns with stratigraphic significance: a progradational unit (A) showing a sigmoidal pattern (see chapter 5); a wedge of sediments (C) that is missing from the top of the structure, indicating that the structure was positive and probably growing at the time of their deposition; and an erosional channel (D), possibly the result of channeling around the positive feature to the left. Color displays of amplitude and frequency aid in defining these features (see color plate 8). Much of the growth history can be worked out from seismic patterns, and in this instance the history is critical with respect to hydrocarbon accumulation. The structure, incidentally, shows a bright spot and produces gas from carbonate porosity.

The nature of most rocks varies laterally; for example, sand grades into shale, or lime content increases in a gradual facies change. As this occurs, we expect the rock properties to change and, therefore, the details of a reflection from an interface bounding this rock member. Examining changes in

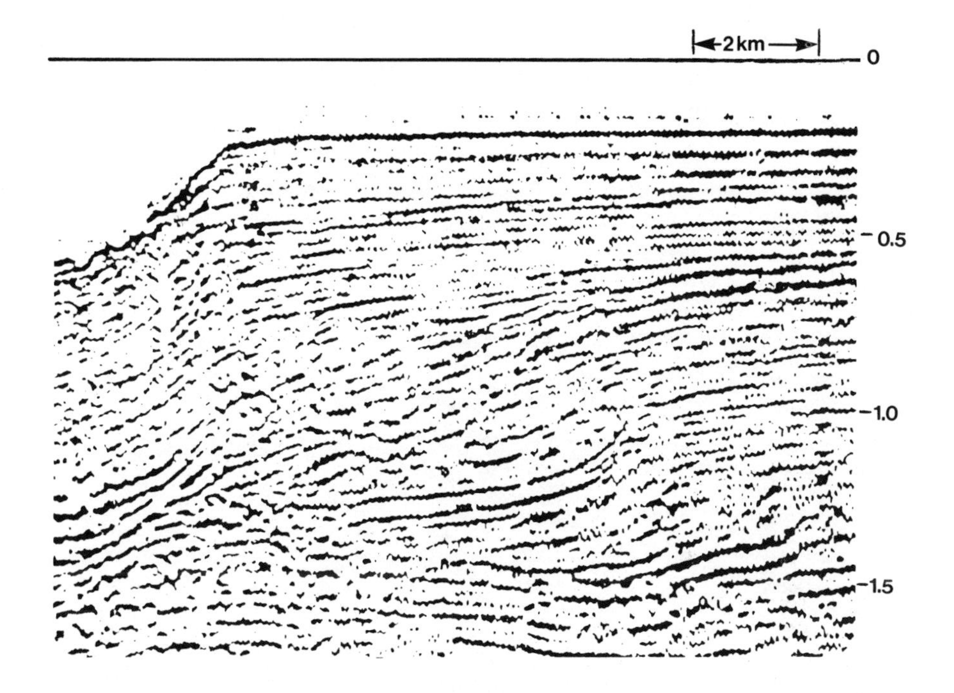

Figure 1.5. Continental slope-shelf break in present seafloor and ancestral shelf edges in the deeper reflections. (Reprinted by permission of United Geophysical Corporation)

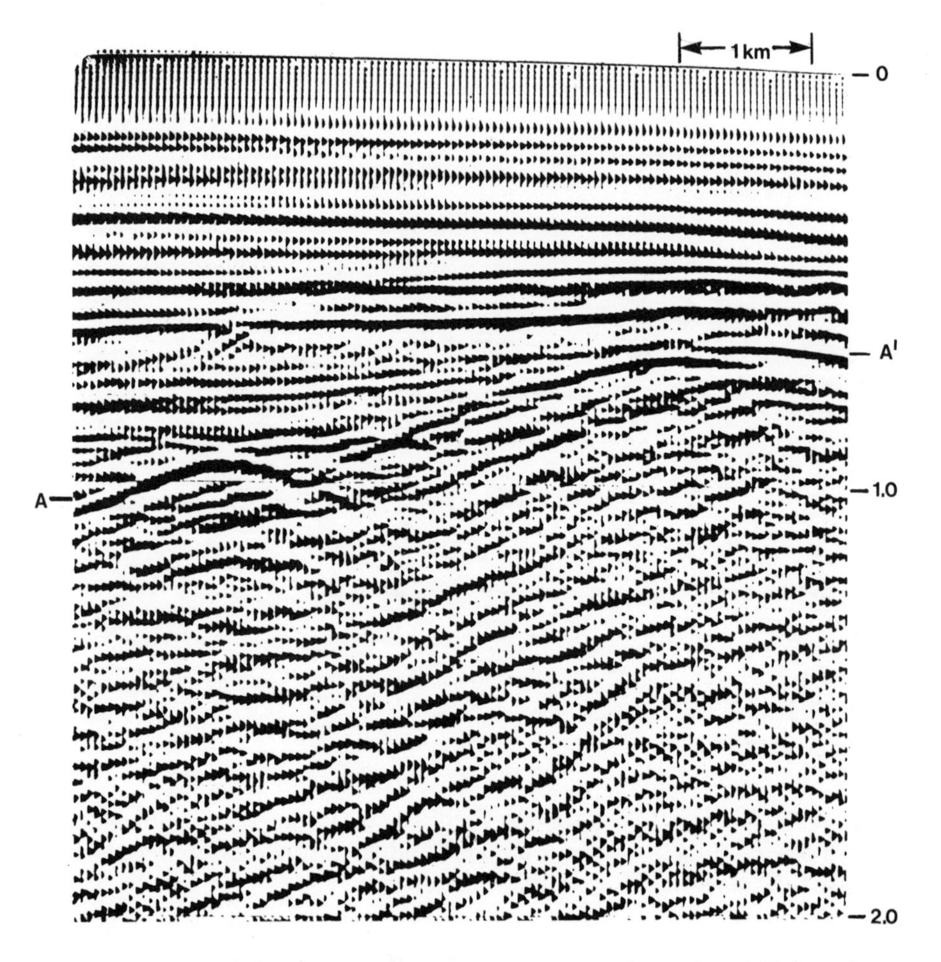

Figure 1.6. Reflection onlap onto an erosional surface (AA′), indicative of a transgression.

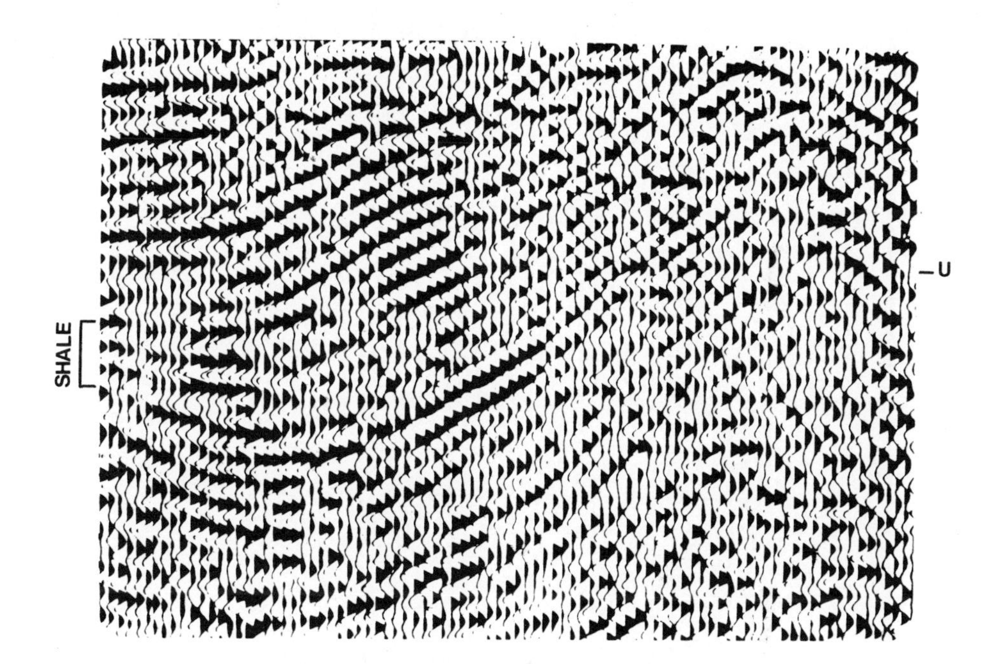

Figure 1.7. Portion of a migrated section over a tight fold. The shale unit varies in thickness through the fold, the competent beds below it do not. U indicates an unconformity.

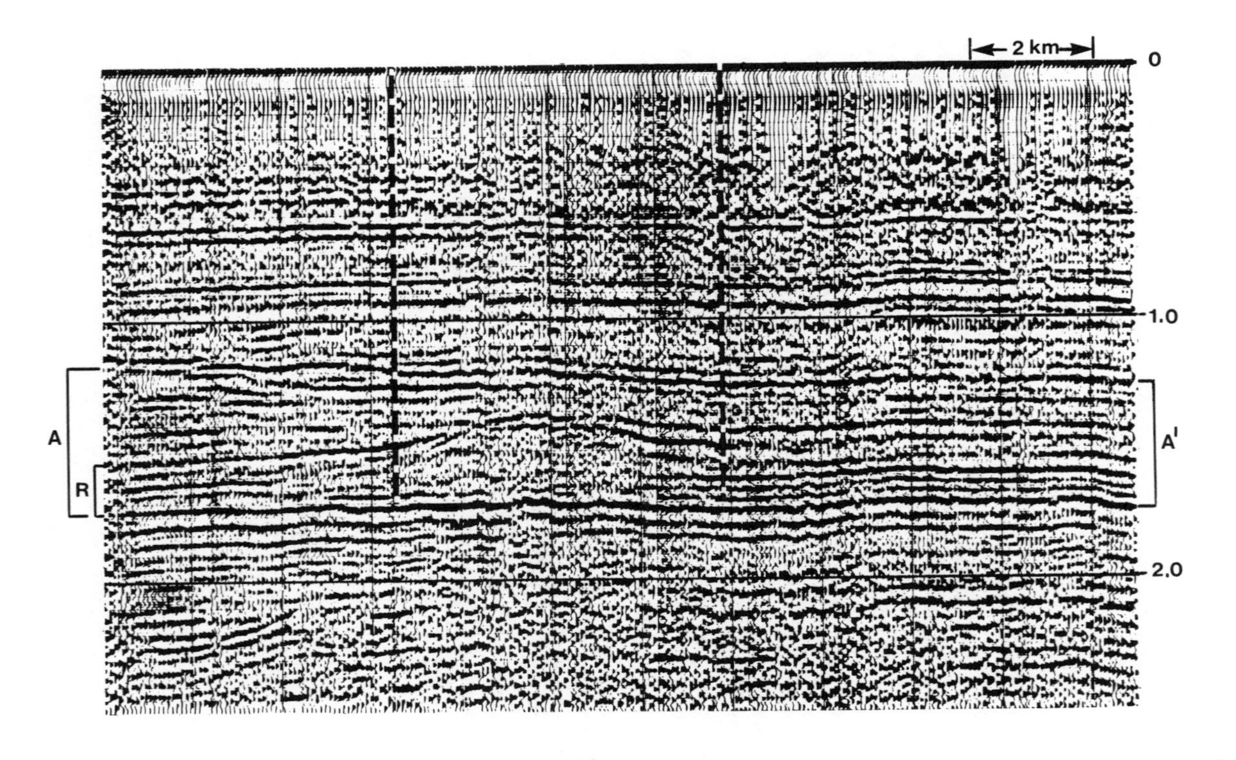

Figure 1.8. Section across Horseshoe Atoll in West Texas. R denotes the portion of the section that contains the reef (just left of center). The back-reef area of flat-lying, strong, continuous reflections is to the right. The fore-reef showing an entirely different progradational reflection pattern is to the left. Note deterioration of data quality below the reef. (Reprinted by permission of Conoco Incorporated)

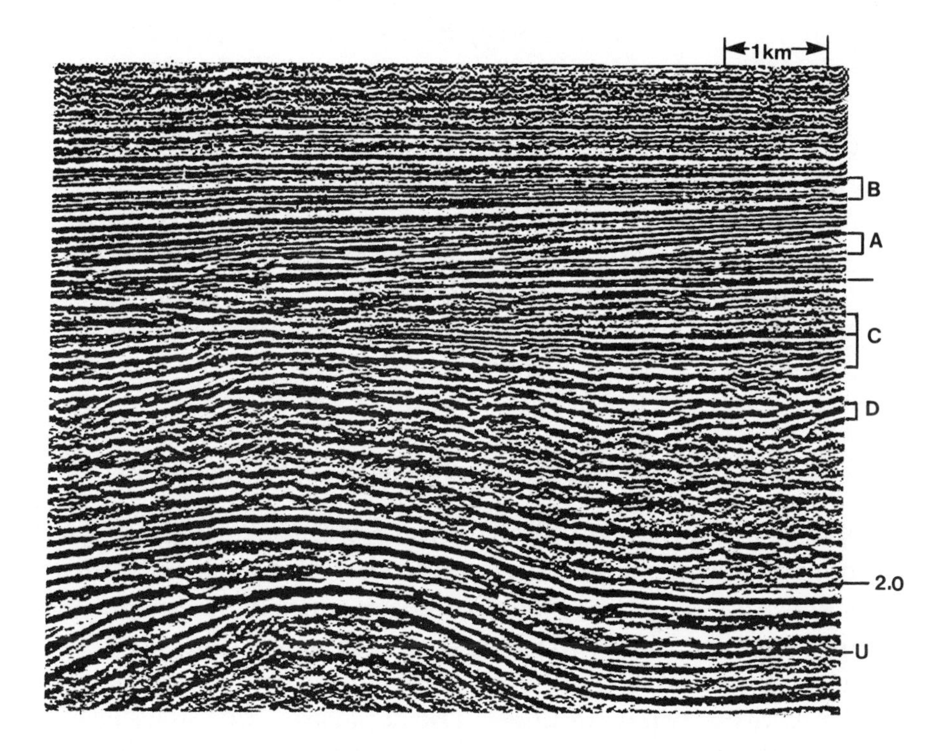

Figure 1.9. Phase section showing several stratigraphic patterns. B is a unit that thickens to the left (note onlap at its base); A shows progradation from a source to the right; C thins onto the structure; D is fill in an ancient erosional channel; and U is an unconformity within a mainly carbonate section. The anticline produces gas from about 1.8 s (see also color plate 8). (Reprinted by permission of Seiscom Delta)

individual reflections or small groups of reflections is called *reflection character analysis* (see chapters 7 and 8). Ancillary measurements made on the data often emphasize character changes, and color displays help localize such changes. The unit AA′ in figure 1.10 involves higher velocity rocks on the right third than on the thicker section to the left, where the unit consists of clastic sands and shales. This change creates changes in the waveshape, and the changes are emphasized by a polarity display.

Figure 1.11 is a portion of a phase display across a turbidite build-up. Phase emphasizes relatively weak reflections such as those from turbidities. The flat event at A is probably a flat spot from a gas-water contact; flat spots are one of the best hydrocarbon indicators (see chapter 9).

Seismic Stratigraphy Subdivisions

Seismic sections can often be subdivided into units that have common characteristics but that differ from adjacent units. The separation of seismic data into such units is the branch of seismic stratigraphy called *seismic sequence analysis* (see chapters 3 and 4). Figure 1.12 almost naturally subdivides itself into these separate units. The separate units are distinguished by different reflection characteristics, that is, different *seismic facies*, the indicators of the depositional environment (see chapter 5).

The belt of oblique reflection patterns across the center of figure 1.13 separates regular, fairly smooth reflection events above the belt from very irregular reflections below. The velocity of the sediments that constitute the oblique portion varies appreciably, and the variable travel time through this unit creates undulations and other irregularities in the deeper reflections. Higher velocity distinguishes the portions of the oblique reflection that are predominantly carbonate (see chapter 6).

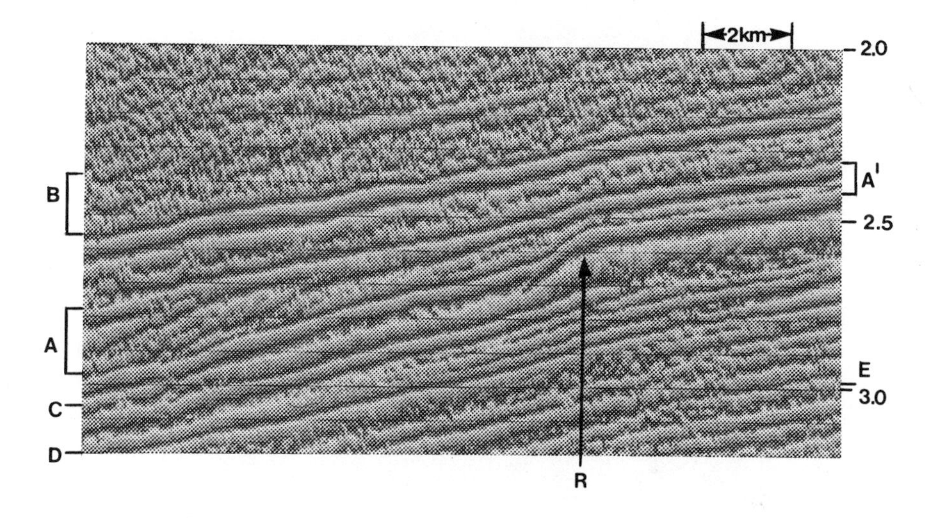

Figure 1.10. The lithology of the unit AA′ changes at the shelf edge, over a reef (R). This is a phase display. There is gas production from the pinchout of sands at the top of unit A, which shows an oblique pattern (see chapter 5). Unit B shows a sigmoidal pattern and its sediments are generally fine grained. C, D and E are unconformities, as well as the top and base of units A and B. (Reprinted by permission of Seiscom Delta)

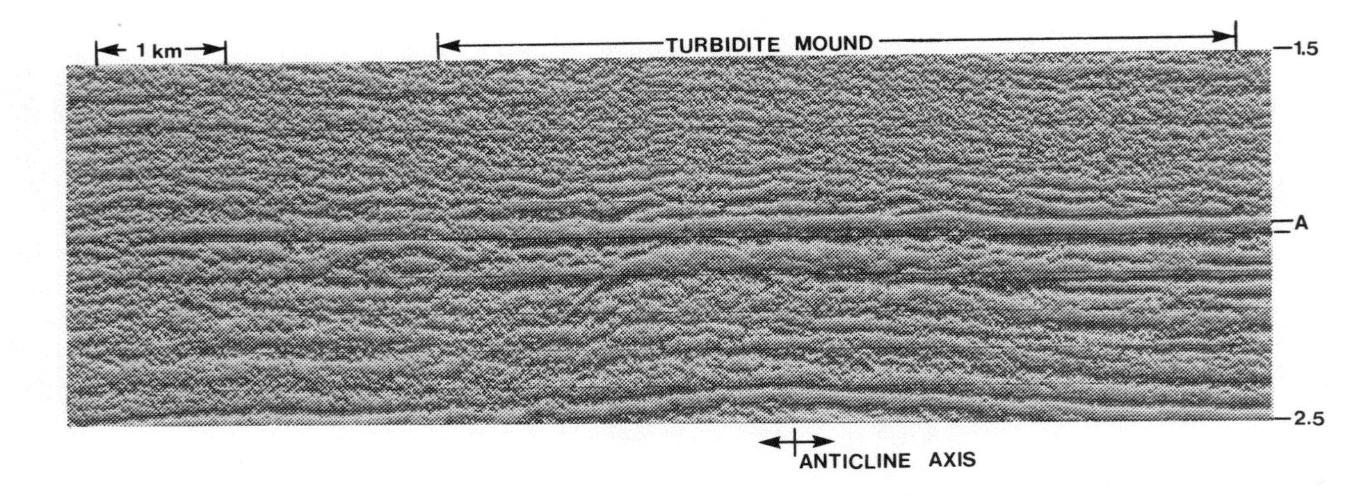

Figure 1.11. Phase section across a turbidite build-up. The event at A is a flat-spot reflection from a gas/water contact. The anticlinal reversal (about 2.4 s) below the turbidites must be real because velocity variations due to the turbidites could not be great enough to produce a fictitious anomaly of this magnitude. (Reprinted by permission of Seiscom Delta)

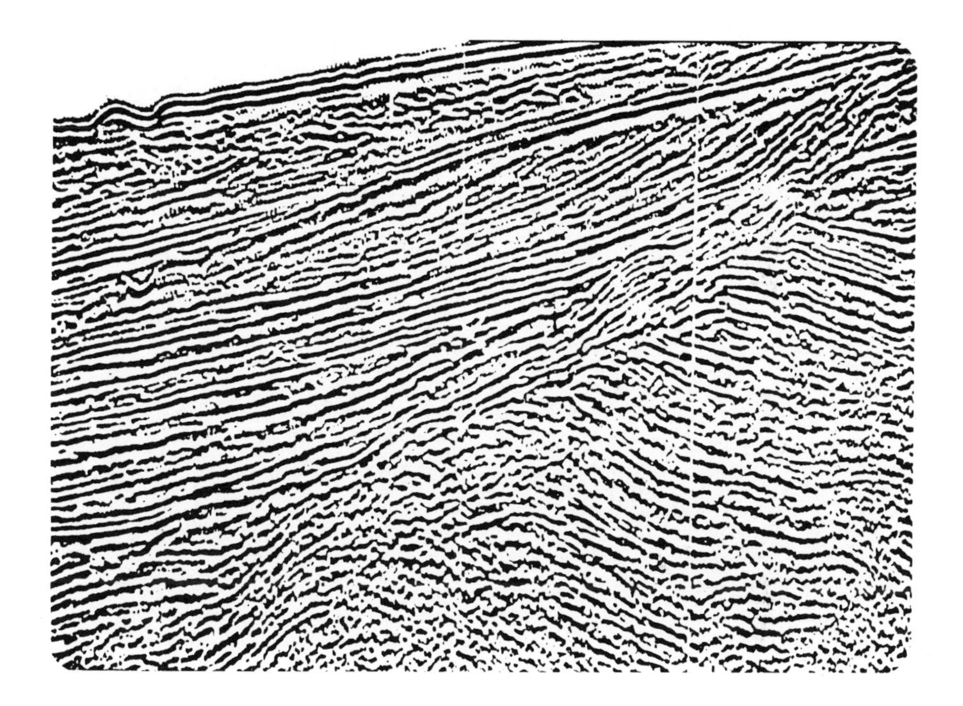

Figure 1.12. Changes in reflection pattern allow the seismic section to be separated into seismic sequence units that differ in reflection character. (Reprinted by permission of Exxon Production Research Company and The American Association of Petroleum Geologists)

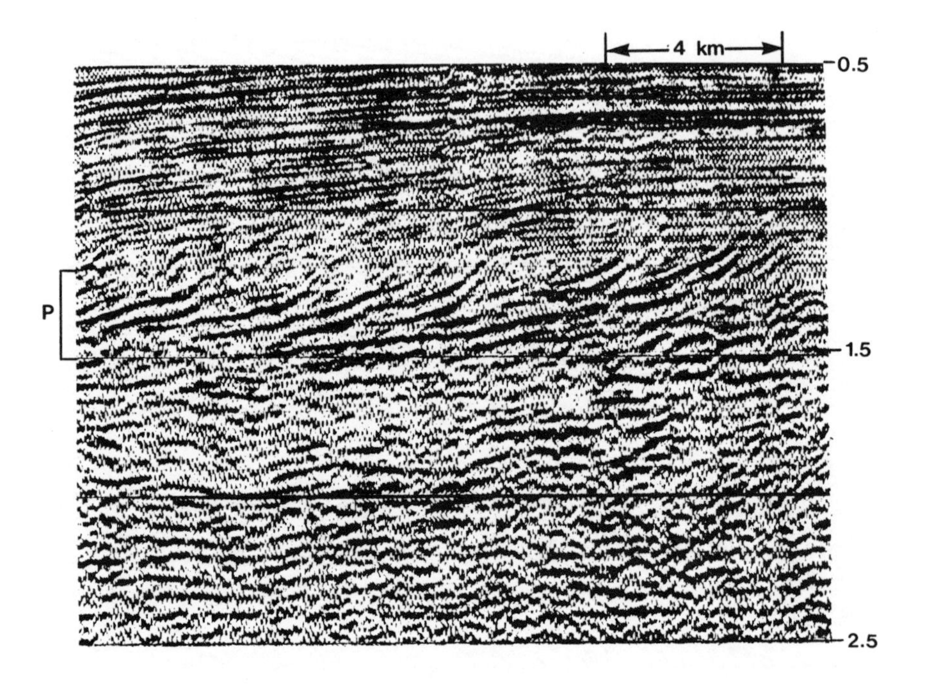

Figure 1.13. The belt of oblique reflections (P) has produced velocity anomalies in the deeper section. (Reprinted by permission of United Geophysical Corporation)

Seismic sequence analysis (along with seismic facies analysis) and reflection character analysis represent the major techniques employed in seismic stratigraphy. The interpretation procedure is illustrated in figure 1.14. Seismic sequence analysis isolates depositional units. Seismic facies analysis is generally concerned with determining the depositional environment and lithofacies, concentrating on the updip edges of units where depositional energy was probably greatest. Working out the geologic history is one of the key objectives of seismic stratigraphy. Reflection character analysis concentrates on changes involving single (or a few) reflections and endeavors to interpret these in terms of stratigraphic changes or hydrocarbon accumulations. It thus is involved in predicting reservoirs and traps.

Seismic Limitations

Improved seismic data quality has made stratigraphic interpretation from seismic observations possible. However, seismic interpretation is subject to a number of important limitations. Of special importance are those involving (1) resolution both vertical and horizontal, (2) wavelength or bandpass, (3) noise, (4) out-of-the-plane reflections, and (5) velocity variations (see chapters 2 and 6).

Stratigraphic Traps

When estimates are made of the oil and gas remaining to be discovered, that which is trapped stratigraphically (as opposed to structurally) almost always dominates, although different individuals estimate widely varying amounts. The early Seismic Stratigraphy Schools included a session reexamining the case histories of stratigraphic accumulations,

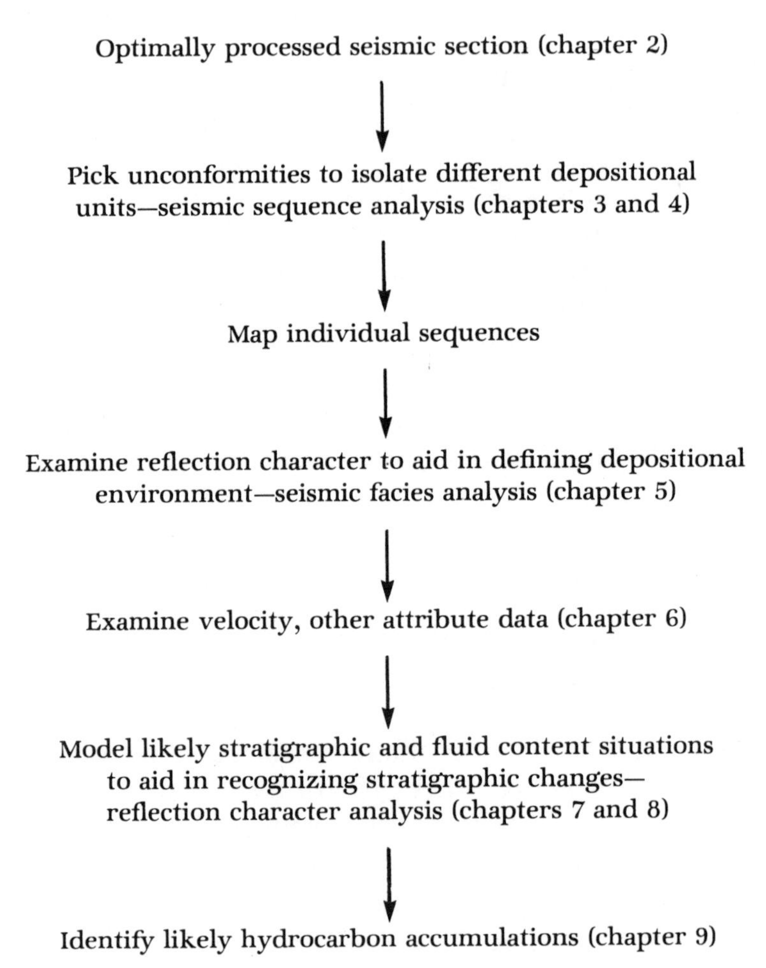

Figure 1.14. Seismic interpretation procedure.

but this portion was dropped from the course because it implied an erroneous conclusion. The picture that the already-documented, rather-old case histories gave was that stratigraphic traps were found by searching for something else—serendipity. We did know of deliberate searches for stratigraphic accumulations which were successful, but generally these were not available for public disclosure. This session of the schools usually ended with a plea to disclose present-day experience.

Nevertheless, the stratigraphic case histories had one important moral. While the discovery of stratigraphic accumulations was not generally attributed to a sound exploration program, the genius lay in being alert when a surprise occurred. Often the surprise occurs in the record from a borehole; some portion differs from what we expected in such a way as to suggest the possibility of a stratigraphic trap nearby. But where? This is where reflection character analysis comes into its power; it can help us locate the nearby accumulation that the unexpected in a well suggests. It can help us search for stratigraphic traps directly rather than relying on luck and statistics.

2
Processing for Stratigraphic Interpretation

The recent prominence of seismic stratigraphy reflects our ability to obtain more information from seismic data. This is largely the result of cumulative improvements in data acquisition and processing techniques. The major improvements have resulted from:

(1) common-depth-point methods
(2) digital recording
(3) computer processing
(4) display improvements

Processing Objectives

Data processing usually has one of four objectives:

(1) The improvement of the signal-to-noise ratio, either by enhancing the signal or attenuating the noise. This is the objective of most seismic data processes, which are generally designed to attenuate specific types of noise.

(2) The repositioning of data elements (migrating). The data on unmigrated seismic sections are referenced to the locations of the source and receivers, usually being plotted at the midpoint between them; migration repositions data to the locations of the reflecting points.

(3) The measurement of "attributes" of the data, including velocity, amplitude, frequency, polarity, and other measurements (see chapter 9).

(4) The display of the data in a manner easily understandable by an interpreter; display parameters include the use of optimum scale for the particular interpretation objectives and displays that combine various types of measurements, such as color displays.

Processing must be faithful to timing, amplitude and waveshape, because measurements of these determine the structure and give us stratigraphic information.

Processing involves *models*, that is, concepts of what happened. If the model used is inappropriate to the specific problem, then the processing could miscarry. We could then defeat our objectives and possibly create fictitious pictures of the geology. Examples of models used in processing are (1) the "surface-consistent model" on which statics analysis is based; (2) the model in which the velocity distribution in the earth can be represented by a series of horizontal parallel layers, used for stacking and velocity-analysis purposes; and (3) the model in which reflections on a single trace will be repeated systematically because of multiples, the basis of predictive deconvolution; and so on for other processes.

Examples of Processing Improvements

Processing usually begins with an analysis of the data. Analysis is used to understand the problems involved and to determine the processing parameters; analysis is also done after processing to check that we are satisfied with what the processing has done.

Examples of processes designed to attenuate noise include (1) dereverberation to correct for multiple bounces in an overlying water layer (figure 2.1); (2) statics correction to remove the effects of near-surface variations (figure 2.2); (3) normal move-out correction and common-depth point stacking based on velocity analyses (figure 2.3); (4) application of two-dimensional (velocity) filtering to attenuate coherent noise wavetrains (figure 2.4); and (5) coherency filtering to improve continuity (figure 2.5). Obviously proper processing can considerably enhance our ability to draw correct stratigraphic conclusions.

The applicability of processes sometimes depends upon the specific objectives of the processing. Someone has to determine the processing parameters; for example, someone must examine velocity analyses and determine which events are to be enhanced and which attenuated. Erroneous choices can result in a false geological picture (figure 2.3). An objective way of determining which choice is correct is not always clear. The more geologic knowledge given to the person making the processing decisions, the better the choices will be. The interpreter should interact with the processor to make certain that objectives have been met in optimal fashion.

Another example of how processing decisions can influence interpretation is illustrated in figure 2.5. While coherency filtering improves the continuity of events and is useful where following events is otherwise difficult, it obscures features such as faulting. If the definition of minor

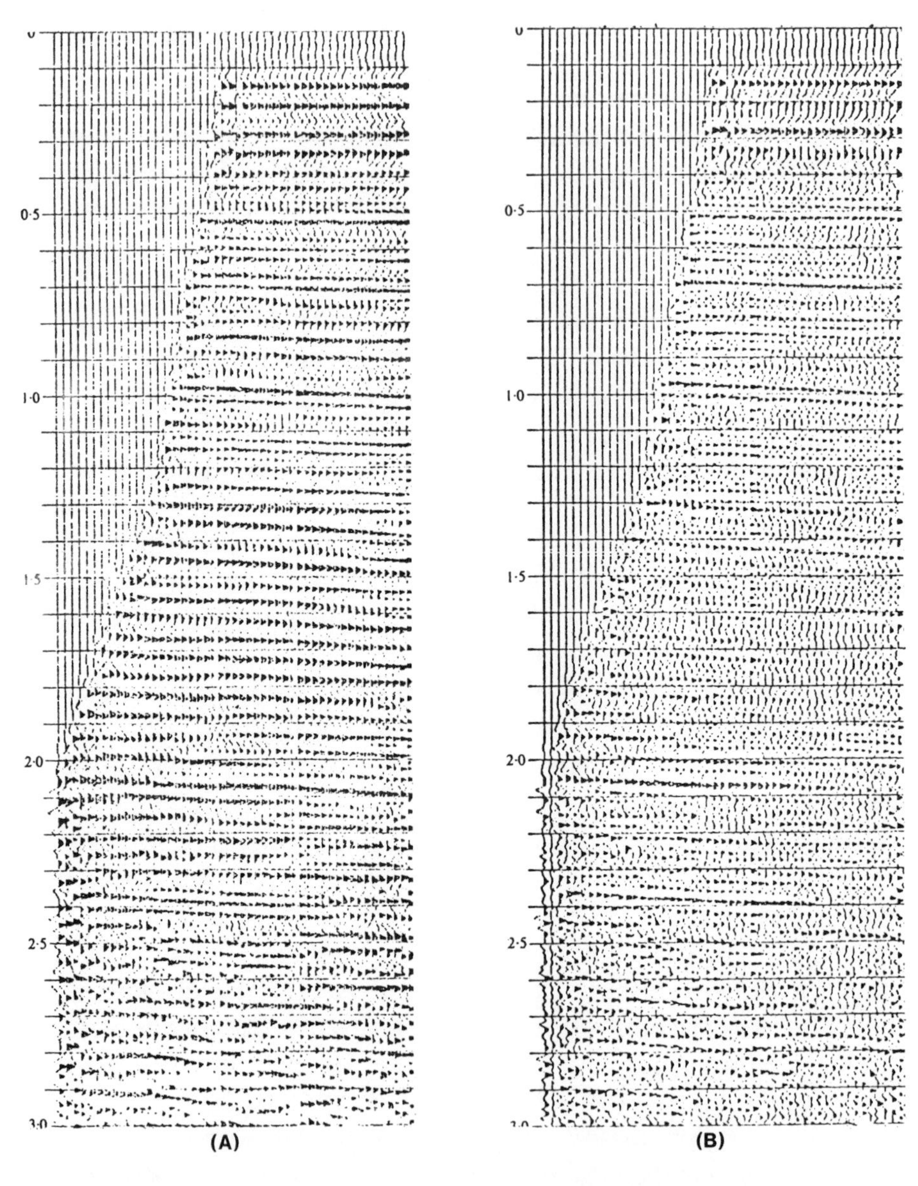

Figure 2.1. Effect of deconvolution to remove water-layer reverberations. (a) Ringy record (also called "singing record") without deconvolution; (b) after deconvolution. (From Telford, Geldart, Sheriff and Keys, 1976; reprinted by permission of the Cambridge University Press and Petty Ray)

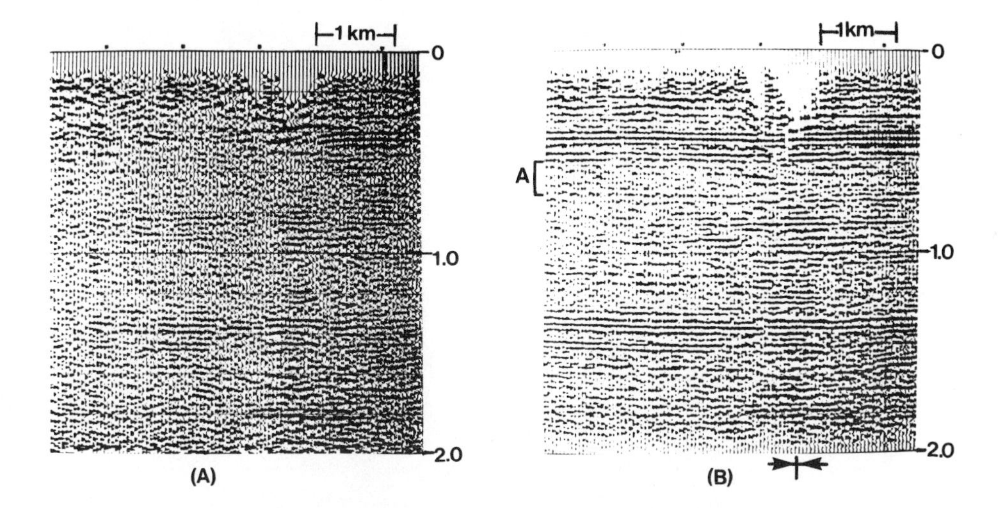

Figure 2.2. Effect of statics corrections: (a) without statics correction and (b) with surface-consistent statics correction. Stratigraphic variations in A are almost impossible to see without the statics correction. Note that the syncline apparently stopped subsiding at about the time the sediments sediments at 1.0 s were being deposited. (Reprinted by permission of Seiscom Delta)

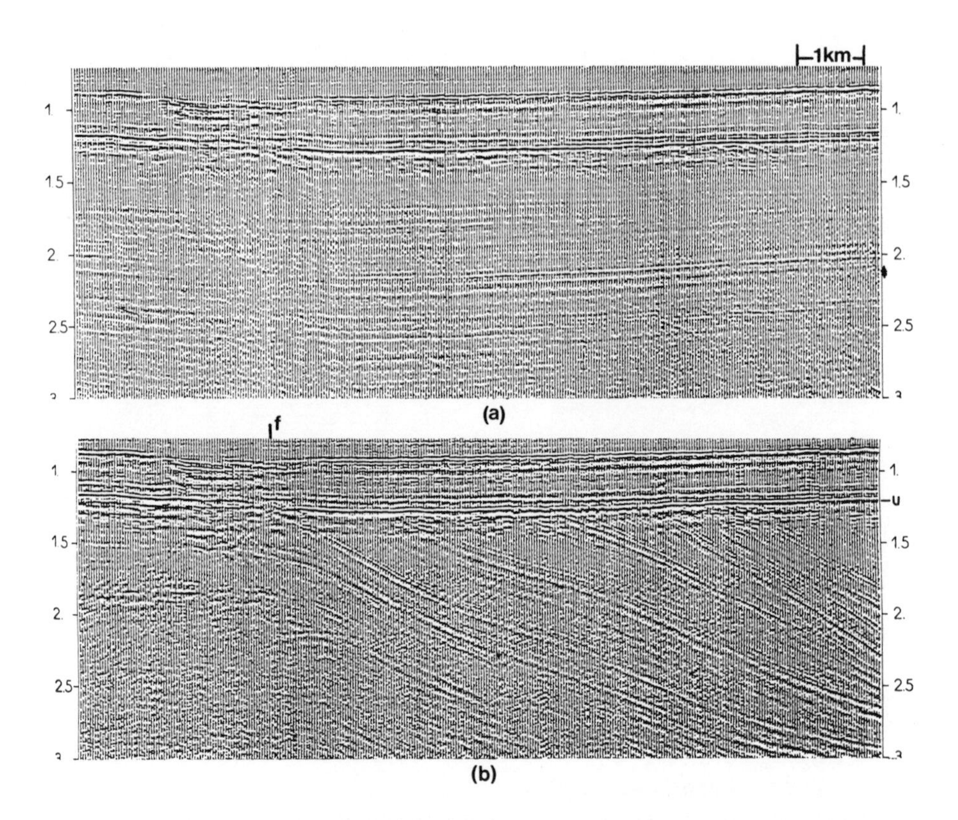

Figure 2.3. Sections assuming incorrect and correct stacking velocities: (a) stacked with velocity that emphasizes multiples and (b) stacked with velocity that emphasizes primary reflections. The unconformity U was the principle generator of the multiples. It is apparently cut at about F by a fault that dips to the right. The exact nature of the fault is not clear, but it has produced disruptions over a sizable portion of the postunconformity section. (From Garotta and Michon, 1967; reprinted by permission of the European Association of Exploration Geophysicists)

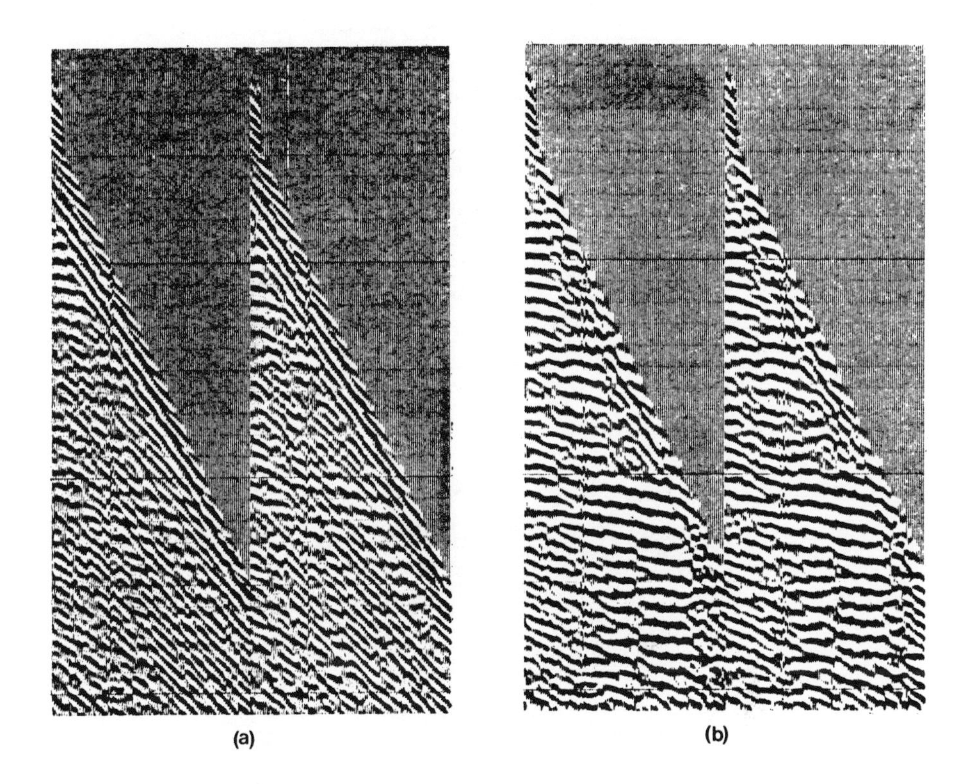

Figure 2.4. Effect of two-dimensional (apparent velocity) filtering in removing shallow refractions: (a) two field records without two-dimensional filtering and (b) after two-dimensional filtering. The two-dimensional filtering permits making more reliable measurements of the curvature of reflection events and hence of the stacking velocity. The two-dimensional filtering also markedly lessens noise interference on the stacked sections. (Reprinted by permission of Seiscom Delta)

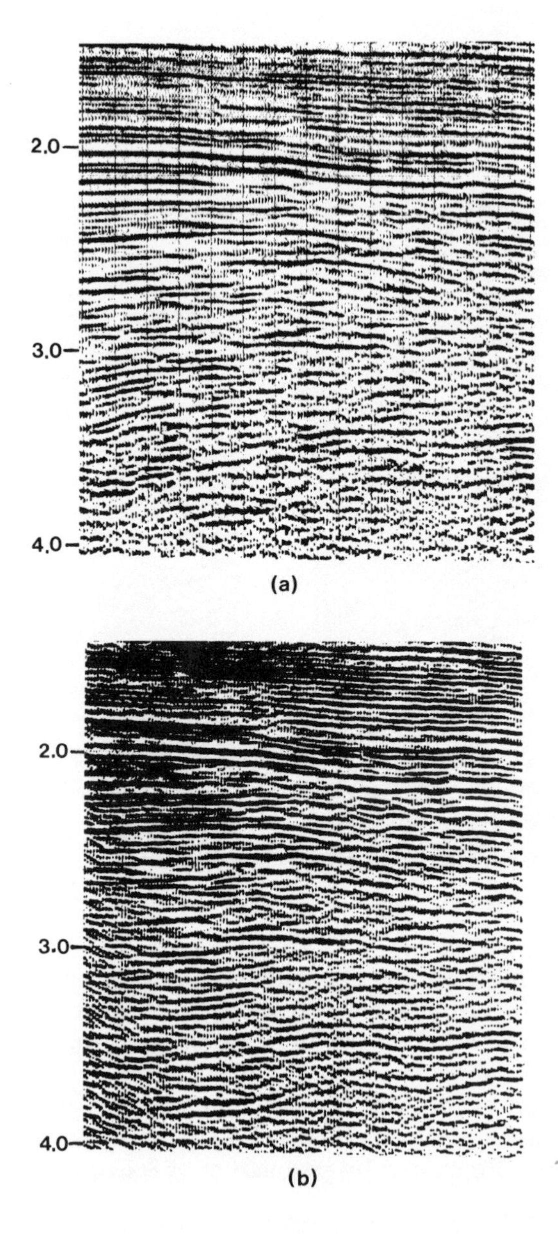

Figure 2.5. Effects of coherency filtering: (a) with coherency filtering designed to improve event continuity and (b) without coherency filtering. The vertical scales and amplitudes of the two sections differ slightly.

faults should be the objective (as it was in this instance), then coherency filtering may make correct interpretation more difficult.

Quality control must include checks to ensure that processing decisions are consistent with interpretational objectives. One type of check is provided by a plot of the velocity that the computer applies in stacking data. Velocity analyses are usually run only every kilometer or so and then interpolated in both time and space by the computer. Occasionally the interpolation will do strange things where there are differences between adjacent velocity analyses. A display of the velocities that the computer applies (figure 2.6) allows the interpreter to check that they are reasonable from a geologic point of view.

Linear inverse modeling (figure 2.7) is another display for control purposes; it shows selected velocity picks, the amount by which data will migrate, and the depth and the velocity of the interval above the event. An interpretation must make geologic sense, which is our first-order criterion for correctness.

Migration

Migration involves repositioning data from a system referenced to the points of observation to a system referenced to the reflecting points. It attempts to position data so as to give an image of the reflector; it is also called *imaging*. Most stratigraphic evidence (for example, evidence of unconformities, onlap, reefing, and so on) involves angularity between two or more reflections, and hence the reflection elements will not be correctly located with respect to each other unless the data are migrated.

Figure 2.8 shows a salt-collapse edge with an outlying pod of residual salt. This is difficult to see on the unmigrated

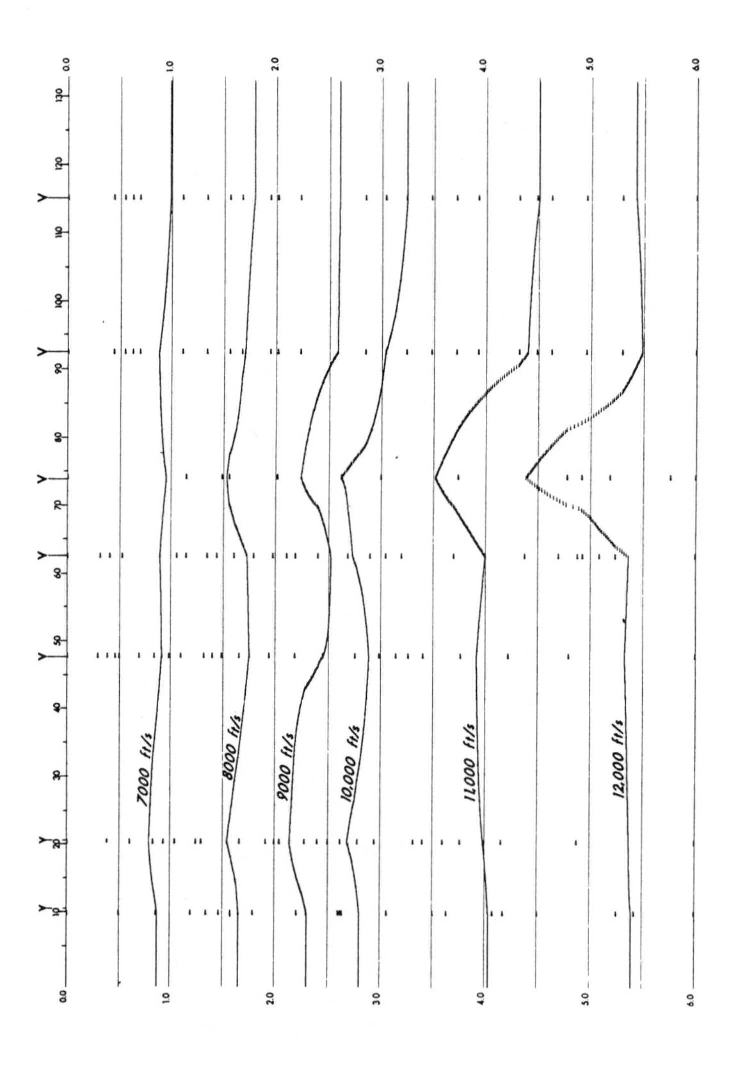

Figure 2.6. Interpolated stacking velocities. Velocity analyses are made at locations indicated by V at top. Times of individual picks are shown by small dashes. Computer interpolates in time and then in space to get complete stacking velocity field indicated by the 1,000 ft/s contours. The velocity analysis at SP 74 is over a salt dome, so the velocity changes in this vicinity are likely real. (Reprinted by permission of Seiscom Delta)

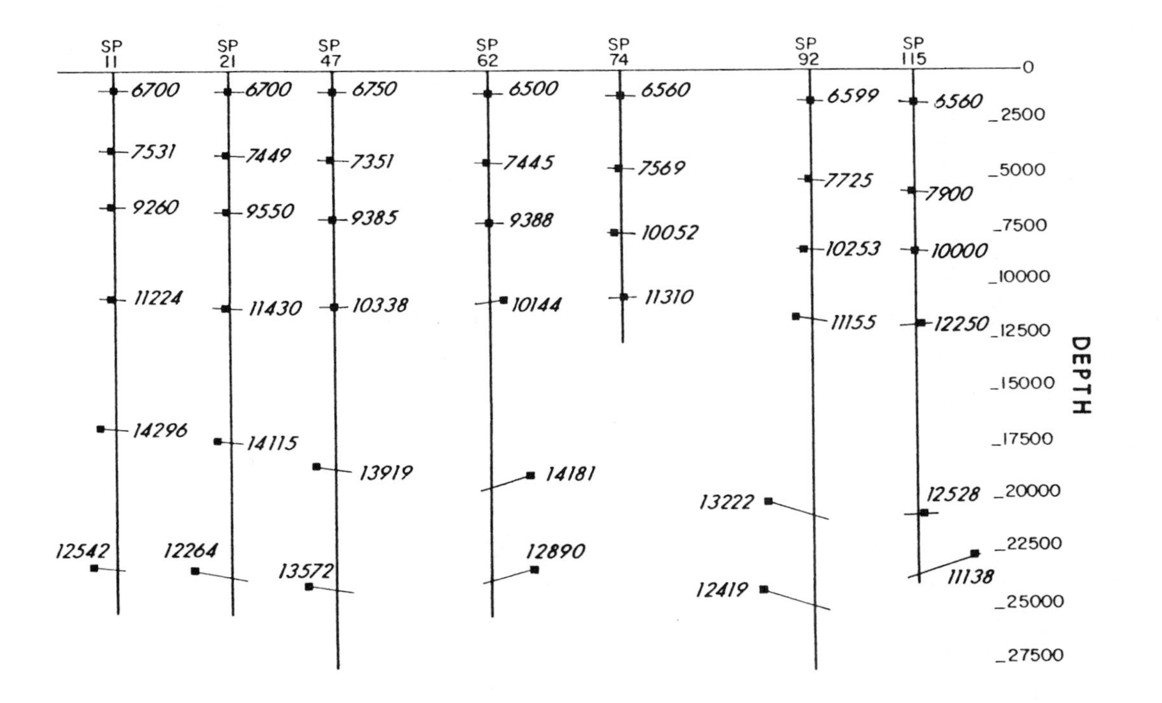

Figure 2.7. Linear-inverse-model display showing locations at which velocity analyses have been run, the migrated positions (small squares) and dip attitudes of reflectors picked on the velocity analyses, and velocity of the interval above the reflector. A deep diapiric salt structure underlies the vicinity of SP 74. This display is for the same line shown in figure 2.6. (Reprinted by permission of Seiscom Delta)

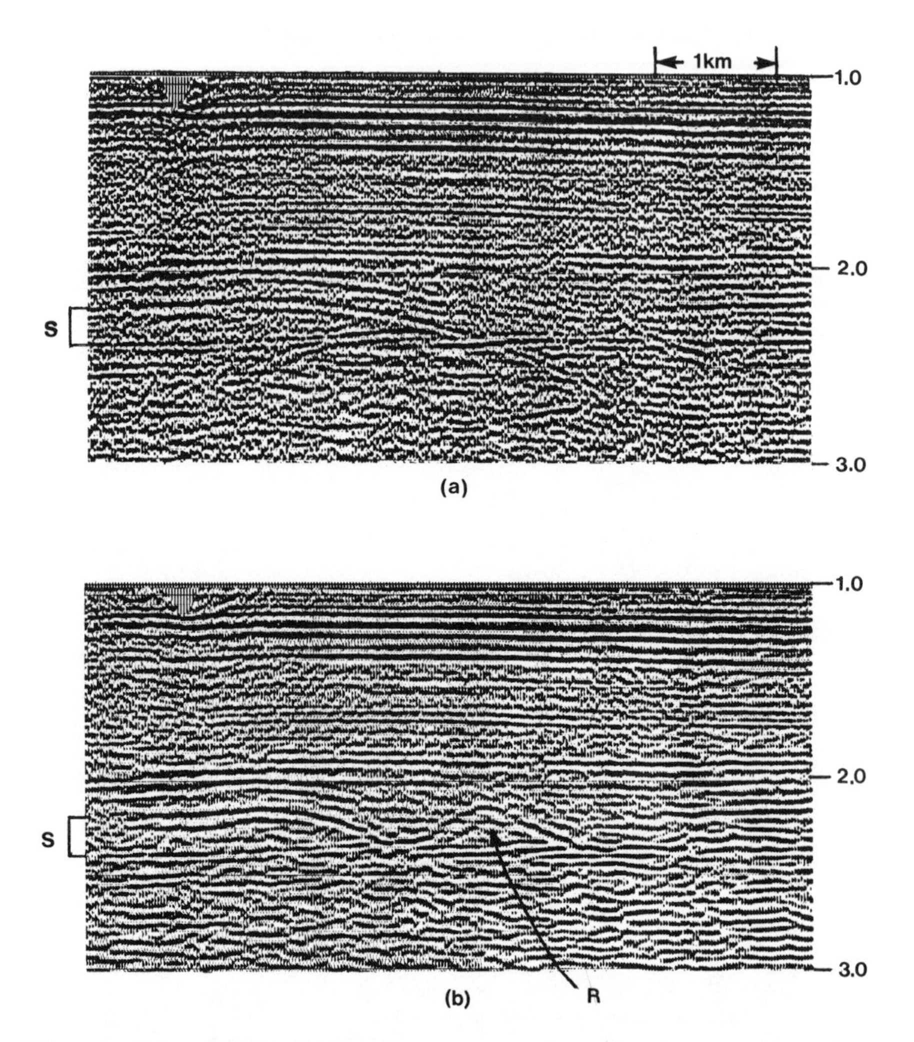

Figure 2.8. Effect of migration on a stratigraphic feature: (a) without migration and (b) with wave-equation migration. The zone marked S is a salt member that has been removed by solution over part of the region. Subsidence during the salt removal allowed slight thickness increases in the units being deposited at that time, so variations in thickness of overlying units provides a method for localizing the salt-solution zones and determining the time of salt solution. The pod of residual salt R in front of the main salt-solution edge could have been missed easily on the unmigrated data. (Reprinted by permission of Seiscom Delta)

data, but it is clear after migration. Isopach maps showing the time intervals between reflection events provide indications for several types of stratigrapic features. The sediments that were being deposited when the salt was being dissolved are thicker because the sediments collapsed into the void created by the salt removal. This created a depression into which additional sediments were deposited. Significant isopach variations sometimes involve only a few milliseconds.

Figure 2.9 shows another example of improvement in interpretability resulting from migration. Note how much difference in location is involved, even for features involving only gentle dip.

Wavelet Processing

Wavelet processing is applied to several different types of processes, but usually with one of three different kinds of objectives:

(1) Equalizing the wavelet, that is, making the wavelet shape the same everywhere

(2) Replacing the effective wavelet with some other wavelet, usually a zero-phase one

(3) Separating the reflectivity of the earth from the seismic wavelet

Wavelet processing ordinarily begins with determining the effective wavelet shape, which is usually the critical and difficult aspect of wavelet processing. Once this is done, processing is the fairly straight-forward procedure of determining an operator to change the effective wavelet to some "desired" wavelet. The desired wavelet is usually a very short one, hence the processing is sometimes called "wavelet

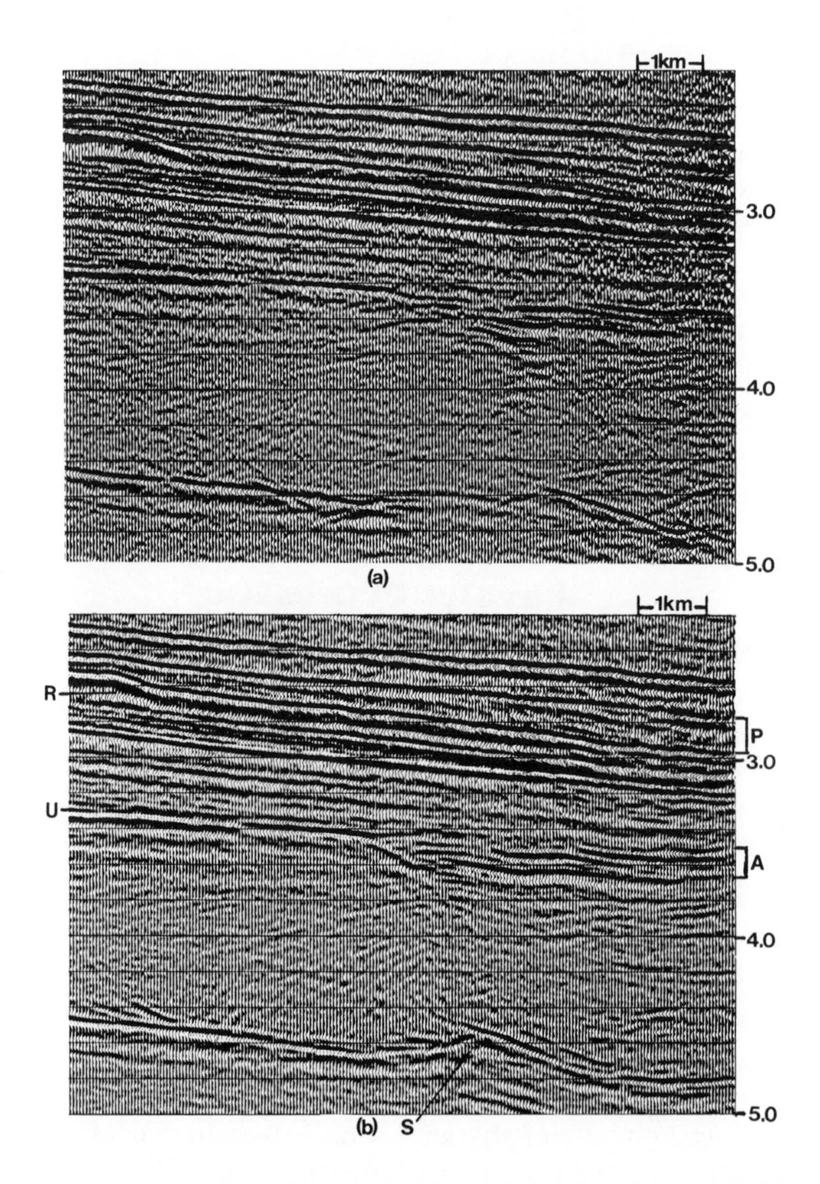

Figure 2.9. Effects of migration: (a) unmigrated section and (b) migrated section. A = high-velocity section not present on left half of section; P = progradational unit; R = reef (note how migration has affected its location); U = unconformity (note downlap pattern onto it); and S = salt uplift. (Reprinted by permission of Seiscom Delta)

compression." An uncertainty limits how much the wavelet can be compressed. The width in time and the width in frequency have a reciprocal relationship: a narrow wavelet requires a broad frequency spectrum, and a narrow spectrum means a wavelet spread out in time. Usually a bandwidth less than 1½ octaves is inadequate and a bandwidth of 2½ octaves or more is desirable.

Wavelet processing in order to equalize the effective wavelet is often done in a prestack mode; such processing is also called *signature correction*. The objective is to make the effective wavelet shape the same for all the traces that are to be stacked together. If the waveshapes differ, the low-frequency components tend to be stacked in phase but the high-frequency components are apt to be stacked out of phase and therefore attenuated. The effect of waveshape variations in stacking is that of a high-frequency filter that narrows the bandwidth and limits the resolution.

Sometimes the source waveform is recorded in the field, in which case a deterministic deconvolution can be used (figure 2.10). More commonly however (especially with land data), the waveform has to be determined by statistical analysis of the data. Most statistical analysis procedures (such as those involving Weiner filtering) assume minimum phase. While most impulsive sources are nearly minimum phase, a few non-minimum-phase roots are usually involved. Better results can be achieved if minimum phase does not have to be assumed. Figure 2.11 shows the source waveforms for three different locations along the same land line before and after prestack wavelet processing; the particular process used determines the wavelet's phase relation and requires only that the wavelet be causal. The benefits of wavelet stabilization and increase in frequency content are illustrated by figures 2.11 and 2.12. The removal of the effects of changes in source waveshape are important in distinguishing waveshape variations produced by stratigraphic features.

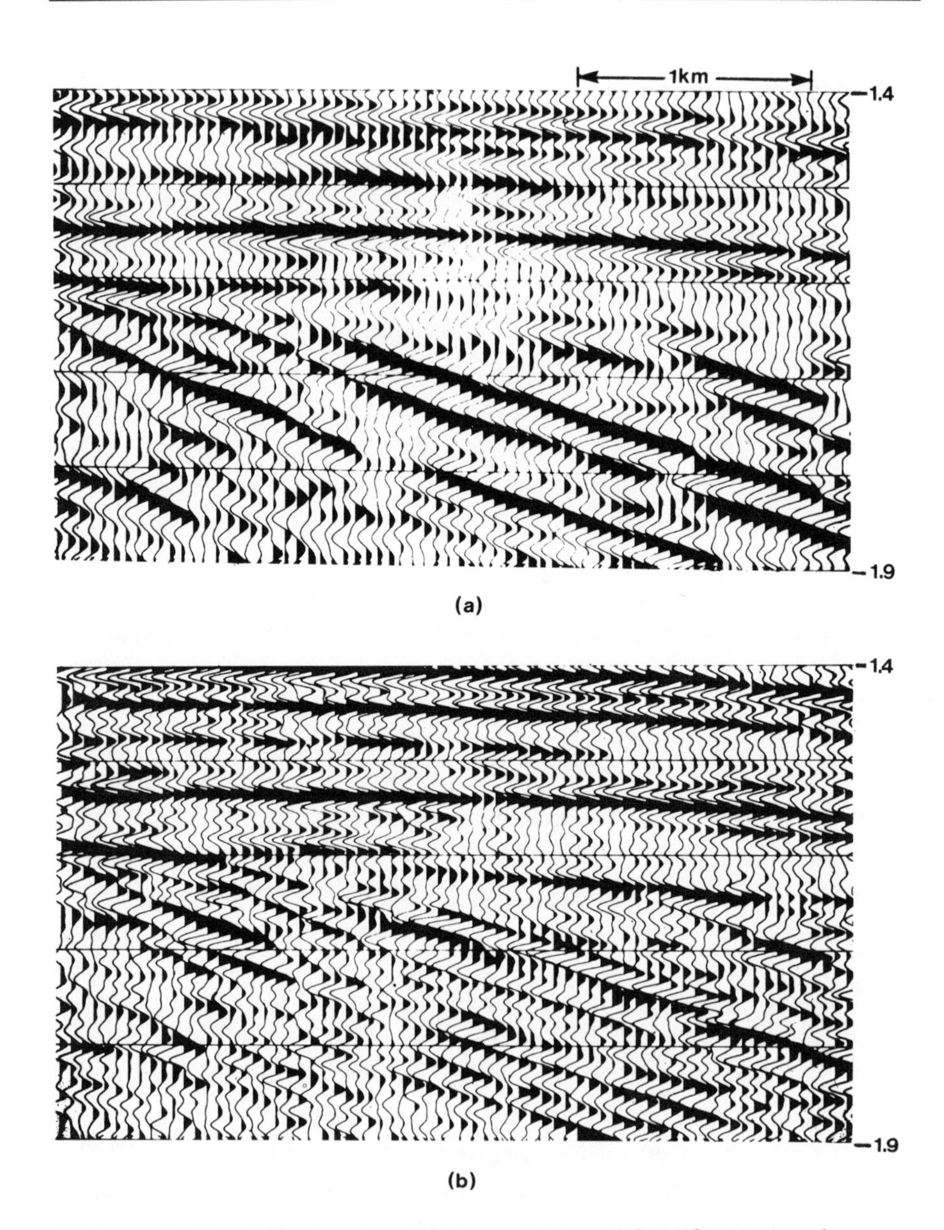

Figure 2.10. Effects of wavelet processing: (a) without wavelet processing and (b) with deterministic waveshape correction based on near-field measurements at each sourcepoint (marine airgun source). (Reprinted by permission of Seiscom Delta)

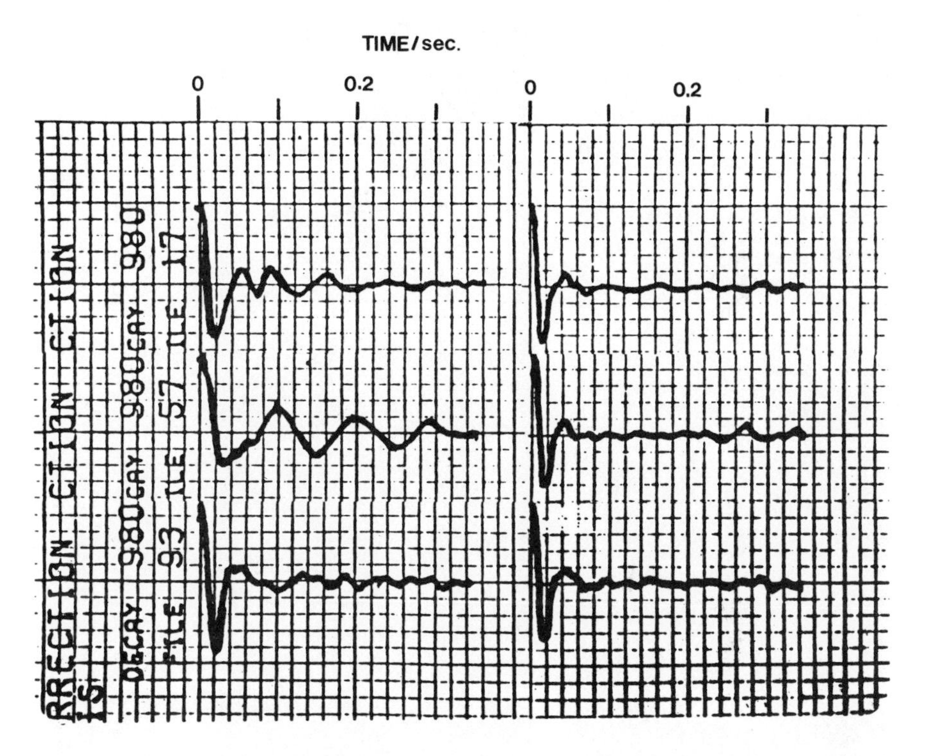

Figure 2.11. Effective source waveforms for three locations on a land seismic line before (left column) and after (right column) wavelet processing of statistical type. (Reprinted by permission of Seiscom Delta)

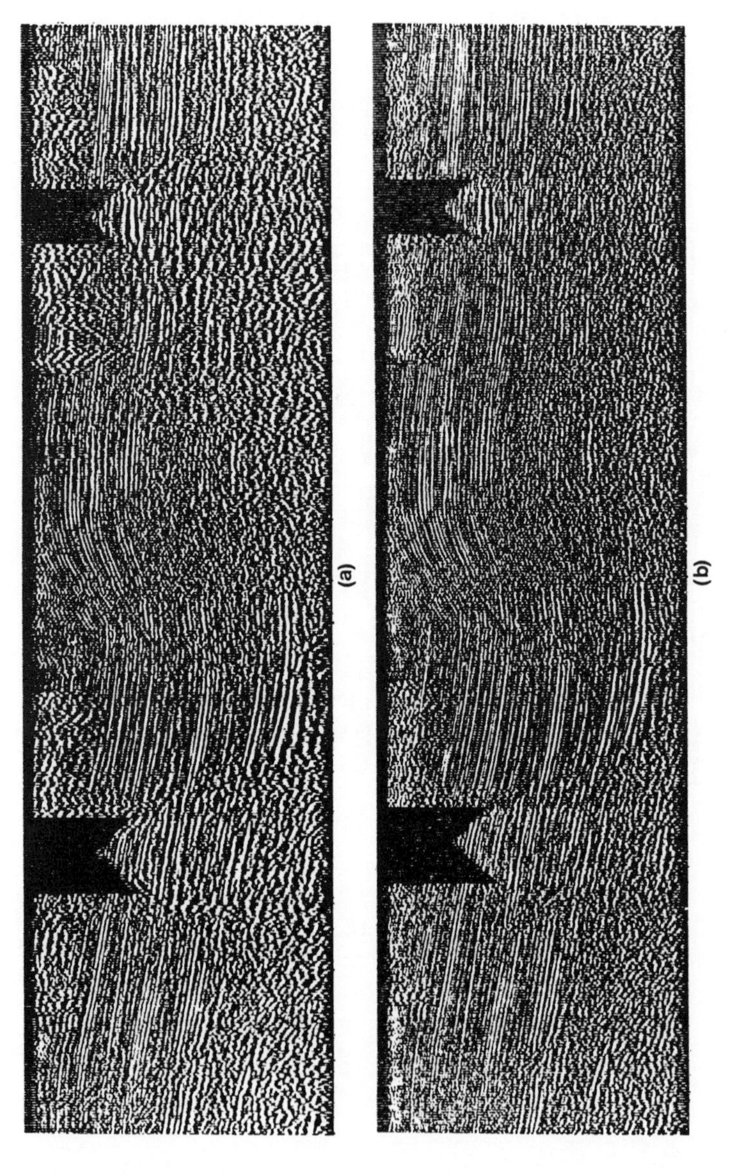

Figure 2.12. Effect of wavelet processing on a land line in an area of appreciable near-surface variation. Undershooting was used to fill in two large gaps on the line. (a) Without wavelet processing and (b) with PULSE prestack wavelet processing. The wavelet processing has lessened waveshape variations, increased the frequency and continuity. A high-angle, basement-involved reverse fault occurs in the middle of the line; this fault probably involves strike-slip motion also. (Reprinted by permission of Seiscom Delta)

Wavelet processing with the object of separating the earth's reflectivity from seismic wavelet effects is based on the convolutional model (see chapter 7). The convolutional model assumes that a seismic trace is simply the result of convolving the earth's reflectivity with an effective wavelet plus additive noise. This type of wavelet processing implies that systematic noise has been completely removed and so usually follows the various noise-attenuating processes, such as statics correction, stacking, deconvolution, and so on, and is usually one of the last processes before display.

Three types of products result from this post-stack wavelet processing: (1) a high-resolution zero-phase section, (2) the earth's reflectivity, and (3) the effective wavelet shape. Each of these may be processed further. The reflectivity may be input to trace inversion in order to simulate acoustic-impedance logs (often called "simulated sonic logs" and "seismic logs"), and analyzed for trace-to-trace variations attributable to stratigraphic changes. The derived wavelet shape may be combined with sonic logs in the manufacture of synthetic seismograms for reflection character prediction studies (see chapters 7 and 8).

Other processes also sometimes called wavelet processing include deterministic processing to remove instrumental filtering effects and source-waveshape corrections that assume that the source was everywhere constant.

Summary

Stratigraphic interpretation requires good field data and good processing. While data processing can improve data quality, it cannot recover information that has been thrown away by poor field recording. Desirable field and processing conditions include uniform conditions along the seismic line; broad bandpass; retention of amplitude significance; good removal of noise; migration; wavelet processing; measuring of attributes; and displays to facilitate mapping the objectives.

3
Seismic Sequence Analysis: The Geologic Models

The stratigraphic model for seismic analysis is one of time-stratigraphic units and the concept that the section can be broken into such units based on seismic observations. To some extent seismic sequence analysis uses duplicate nomenclature—one set from geologic concepts and another from seismic observations. We often identify elements of these different sets with each other and mix up the nomenclature because of this cross-identification. For example, geologic "time-stratigraphic unit" is used interchangeably with "seismic sequence."

A *time-stratigraphic unit* is a three-dimensional set of facies deposited contemporaneously as parts of the same system, genetically linked by depositional processes and environments. The key words here are (1) three-dimensional, (2) contemporaneously, and (3) genetically linked. Contemporaneous depositional systems are illustrated in figure 3.1.

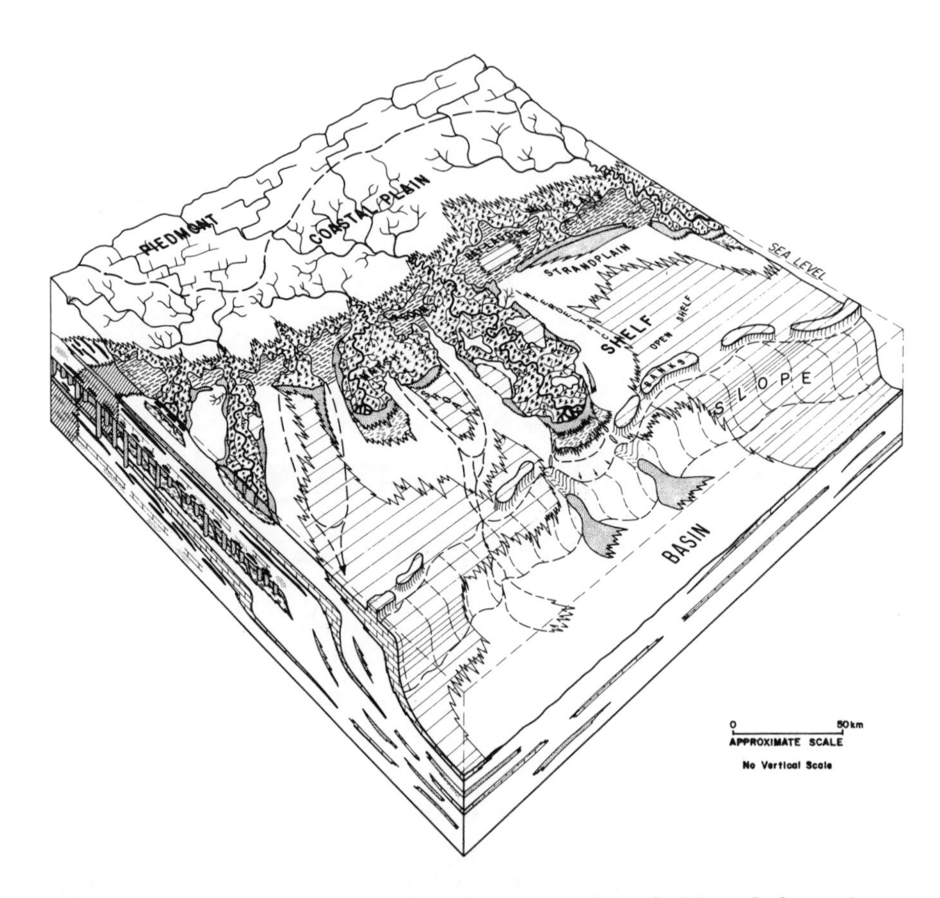

Figure 3.1. Schematic diagram showing delta, shelf, and slope depositional systems operating simultaneously. (From Brown and Fisher, 1977; reprinted by permission of The American Association of Petroleum Geologists)

Different portions of the depositional unit are made of different mixes of grain sizes and lithologies. One might have fluvial sediments and a delta deposited on the landward side of a unit with a carbonate shelf seaward of the delta, reef growth at the edge of the shelf, and still father seaward a fan of sediments on the continental slope and in the ocean basin, all being deposited at the same time and all genetically related to each other. A time-stratigraphic unit is also called a *depositional sequence.*

Geologic Sealevel Change Model

Consider a geologic model to explain the pattern of sedimentation expected from a relative rise of sealevel (because of absolute sealevel rise and/or subsidence of the land surface). As the water rises compared to the land, sediments will be deposited farther and farther landward (figure 3.2) in a pattern called *coastal onlap.* The vertical distance, called *coastal aggradation*, is a measure of the relative rise, allowing for error because of rotation of the overall section and other factors. The associated horizontal distance is called *coastal encroachment.* The nature of the facies that will be deposited at any location depends on both the amount of sealevel rise and the availability of material for deposition. If the terrigenous influx is low, the coast line will move landward with time, producing a *transgression* (figure 3.2a). However, if the terrigenous influx is especially great, the coast line may move seaward with time despite the sealevel rise (a *regression*; see figure 3.2b). Hence, the terms regression and transgression are not by themselves adequate to describe what happened to relative sea level.

In a gradual fall of sealevel (figure 3.3a), successive deposition would be displaced seaward; the tops of the patterns would subsequently be eroded off as they were exposed. If a

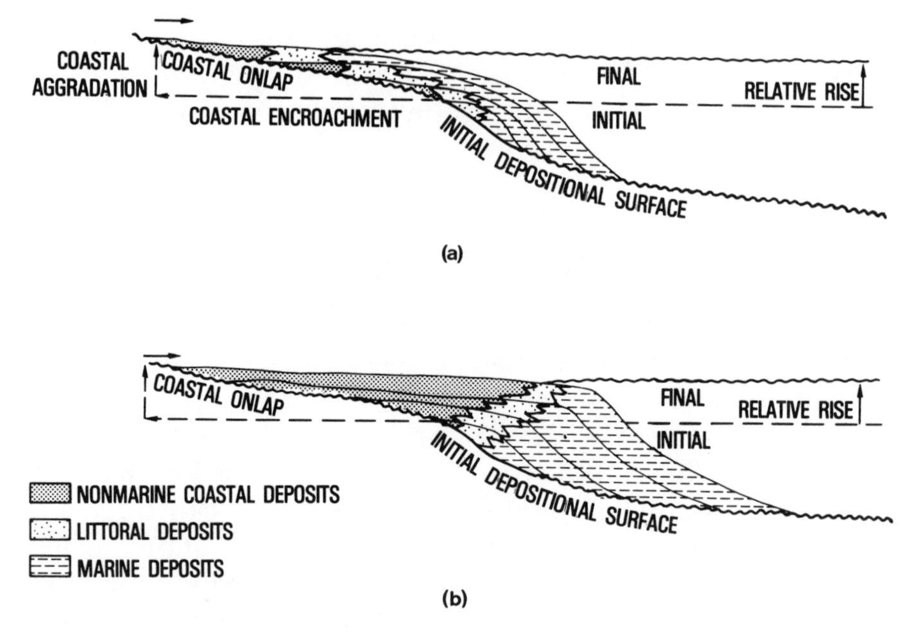

Figure 3.2. Patterns expected for rise in sealevel: (a) if terrigenous influx is low, a transgression results; (b) high terrigenous influx can overwhelm a sealevel rise and produce a regression. (After Vail, Mitchum and Thompson, 1977a; reprinted by permission of The American Association of Petroleum Geologists)

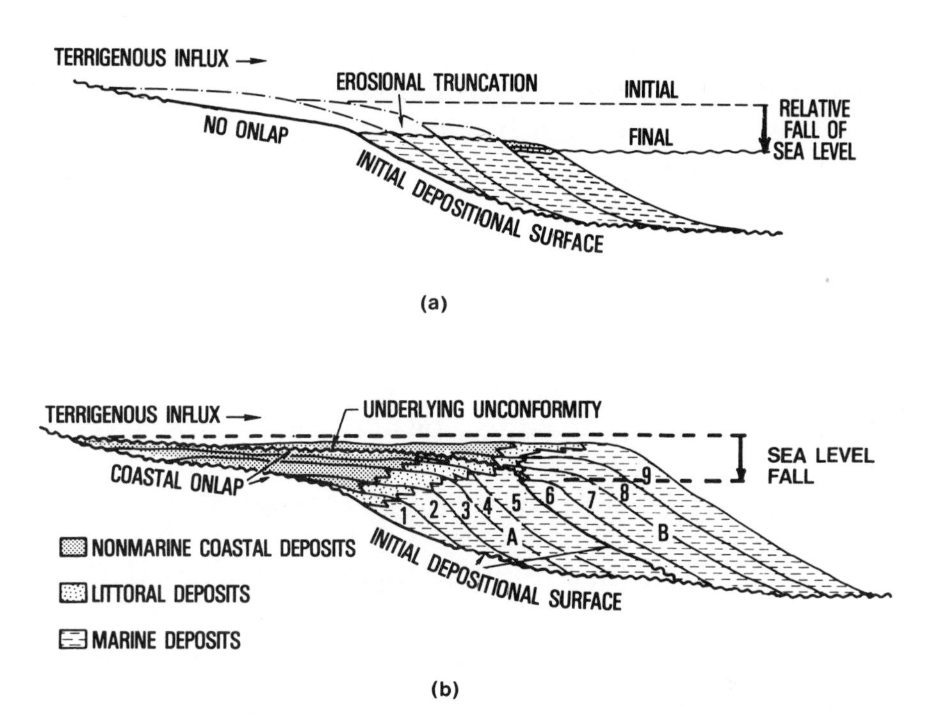

Figure 3.3. Patterns expected for fall in sealevel: (a) a gradual sealevel fall produces a gradual downward shift in depositional patterns, with the tops of the patterns subsequently eroded off and (b) a rapid fall in sealevel produces a major seaward shift in the locale of coastal onlap. The pattern shown is what is expected for long periods of gradual rise interrupted by a short period of rapid fall, between units 5 and 6. (After Vail, Mitchum, and Thompson, 1977a; reprinted by permission of The American Association of Petroleum Geologists)

period of rising sealevel should be followed by a sudden fall, we expect the entire depositional pattern to shift seaward by a considerable amount, giving the pattern indicated in figure 3.3b. The pattern in figure 3.3b is often recognized whereas the pattern of figure 3.3a is almost never seen. The picture therefore is one of asymmetric depositional cycles: a long period of gradual sealevel rise followed by a short period of rapid fall. This is the conclusion of Vail et. al.; an alternative viewpoint is given later in this chapter.

The pattern of gradual rise followed by sudden sealevel fall has presumably been repeated a number of times. There is some evidence to suggest that these patterns in different parts of the world are contemporaneous. This has led to the *eustatic level concept* that periods of rising and falling sealevel represent the same geologic time worldwide (figure 3.4). Eustatic charts reveal the same type of pattern repeated on different scales, which has led to the concept of cycles within cycles (cycles and supercycles). If this pattern is widespread and if we can correlate a local pattern with the master eustatic level chart, then it can be used to age-date sediments.

Reflections as Constant Time Indicators

Seismic reflections generally indicate isochronous (or stratal) surfaces or unconformities. A *constant time* (isochronous or isotime) *surface* at some time was the surface of the solid earth. A time surface does not necessarily indicate the same lithology contrast everywhere; different facies may have been deposited above or below different portions of the surface.

This point is critical to seismic sequence analysis: seismic reflections generally indicate time surfaces rather than the attitude of seismic facies lines, where they differ in attitude. This may appear to contradict expectations because we think of facies lines as separating rock types and, therefore, as of-

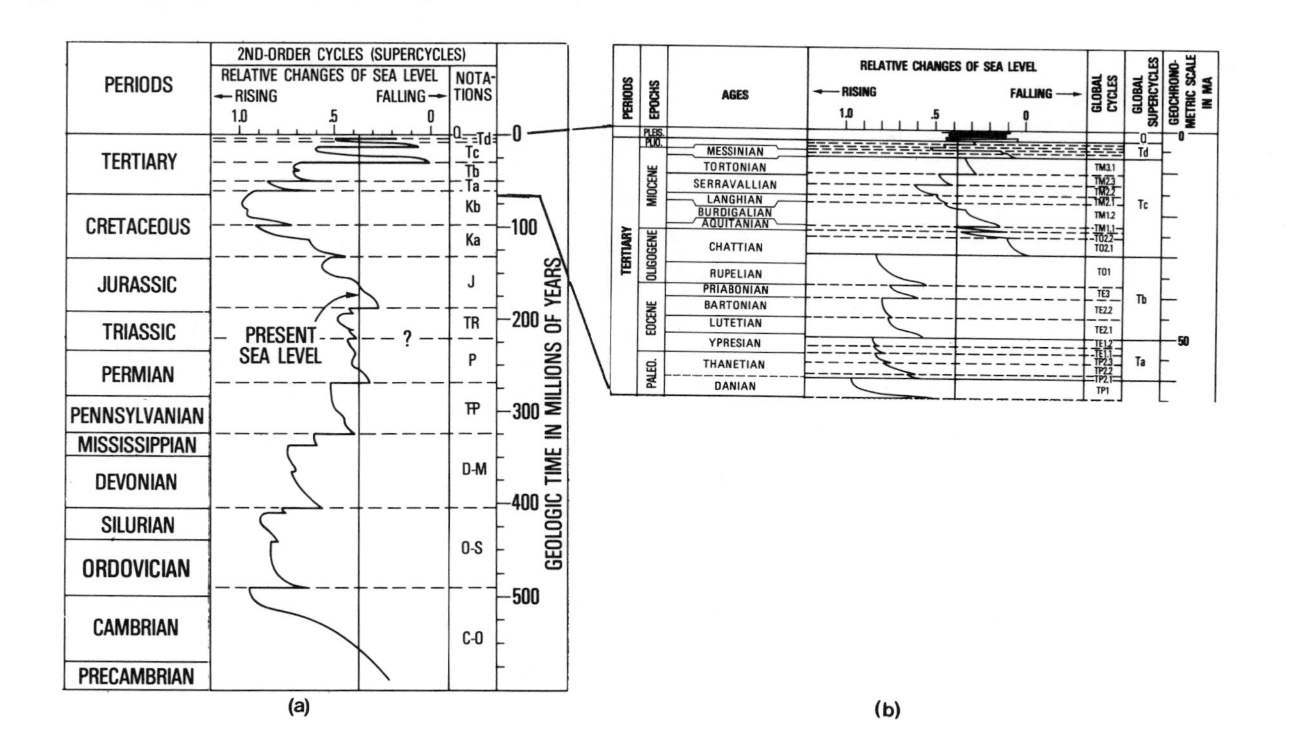

Figure 3.4. Eustatic level charts. (a) Global cycles of sealevel change. Precambrian-to-Triassic and Triassic-to-Recent are the first-order cycles, and the diagram shows the second-order cycles (*supercycles*). (b) Detail of the Tertiary portion of the chart showing the further breakdown into *cycles*. (After Vail, Mitchum, and Thompson, 1977b; reprinted by permission of The American Association of Petroleum Geologists)

fering the contrasts that should produce reflections. The explanation is partly geological, partly geophysical, and partly a matter of sampling density. Let us first examine the geologic considerations.

Geologic events are often episodic, that is, some event such as a storm or flood rearranges the sediments over a large area in a short period of time. The long period of steady deposition between these relatively sudden events may provide the sediment materials, but the sudden events rearrange them and determine the final pattern. These episodic events are often widespread, and the surface that they mark indicates constant time. A succession of such events produces a number of nearly parallel minor but widespread markers.

An example of deposition upon an unconformity is illustrated in figure 3.5. The logs of two wells show a transgressive basal sand laid down on an unconformity; a "top basal sand" line would be a facies line. However, the sand in the right-hand well is younger than that in the left-hand well. With additional intervening control (figure 3.6), it can be seen that time (stratal) lines cross the facies line that marks the top of this basal sand unit. The seismic pattern from this sequence (figure 3.7) shows gradual onlapping at the unconformity, with the seismic events paralleling the time lines rather than the top-of-basal-sand facies line.

Seismic reflections usually are the interference composites of many subreflections. Subreflections result from the often-minor contacts along the time lines. One does not see the individual subreflections because of resolution limitations, but they tend to add in the same way at geophone stations that are not very far apart because the pattern change is so small between adjacent reflecting points. We pick a seismic reflection by deciding what is the same (coherent) from trace-to-trace, and thus we tend to pick as reflections the attitude of the time lines.

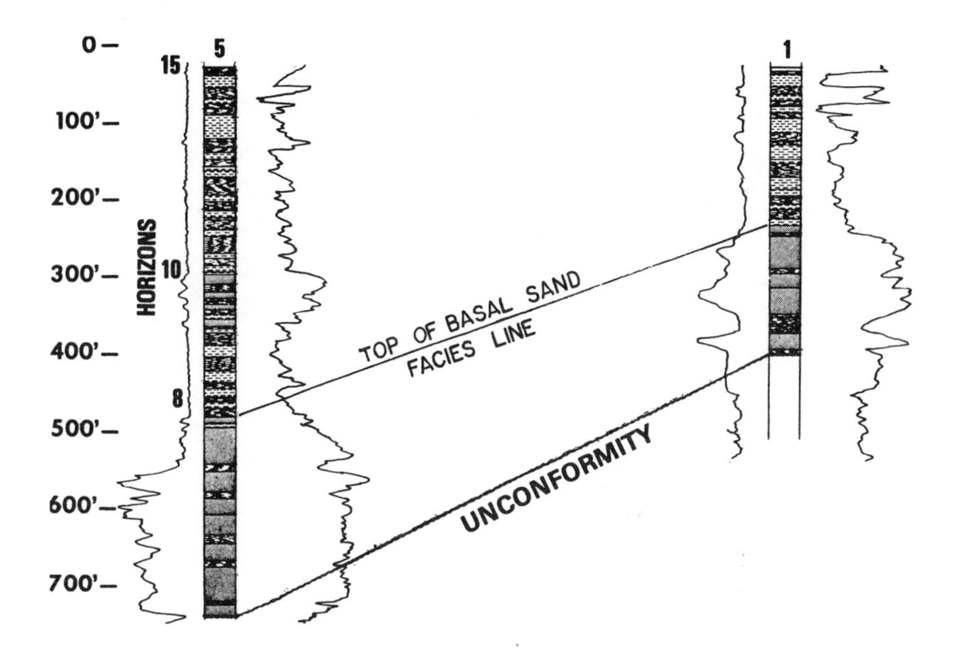

Figure 3.5. Correlation between two wells. (After Vail, Todd, and Sangree, 1977; reprinted by permission of The American Association of Petroleum Geologists)

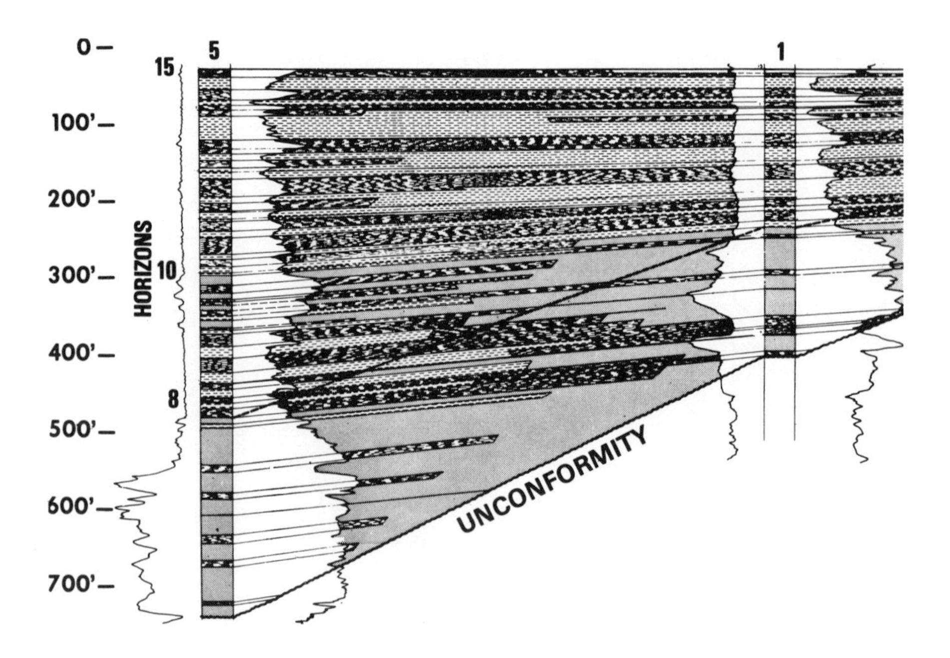

Figure 3.6. Correlation between two wells shown in figure 3.5 after more information is added. A number of other wells intervene between wells 5 and 1. (After Vail, Todd, and Sangree, 1977; reprinted by permission of The American Association of Petroleum Geologists)

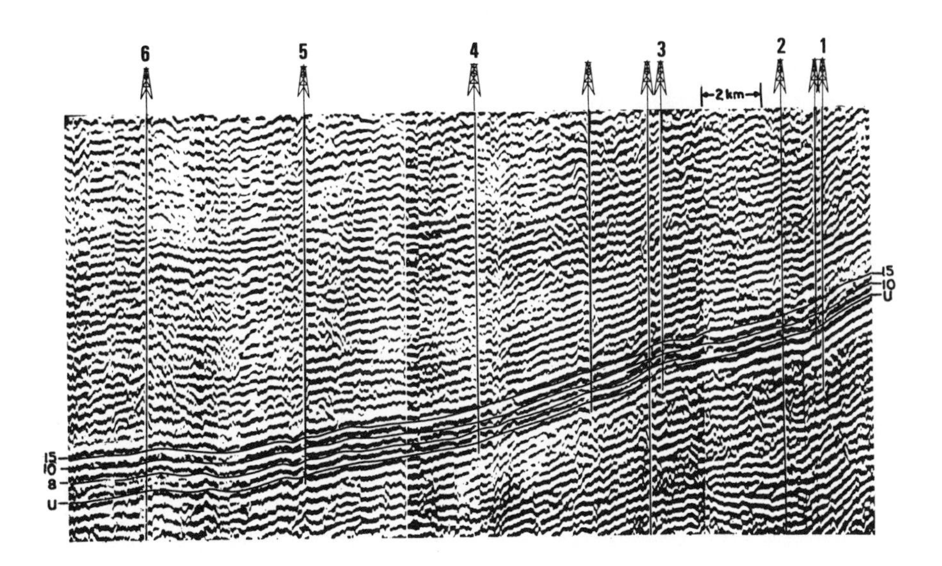

Figure 3.7. Seismic section connecting the wells shown in figures 3.5 and 3.6. U = unconformity; 8, 10, 15 = markers shown in figure 3.6. (From Vail, Todd, and Sangree, 1977; reprinted by permission of The American Association of Petroleum Geologists)

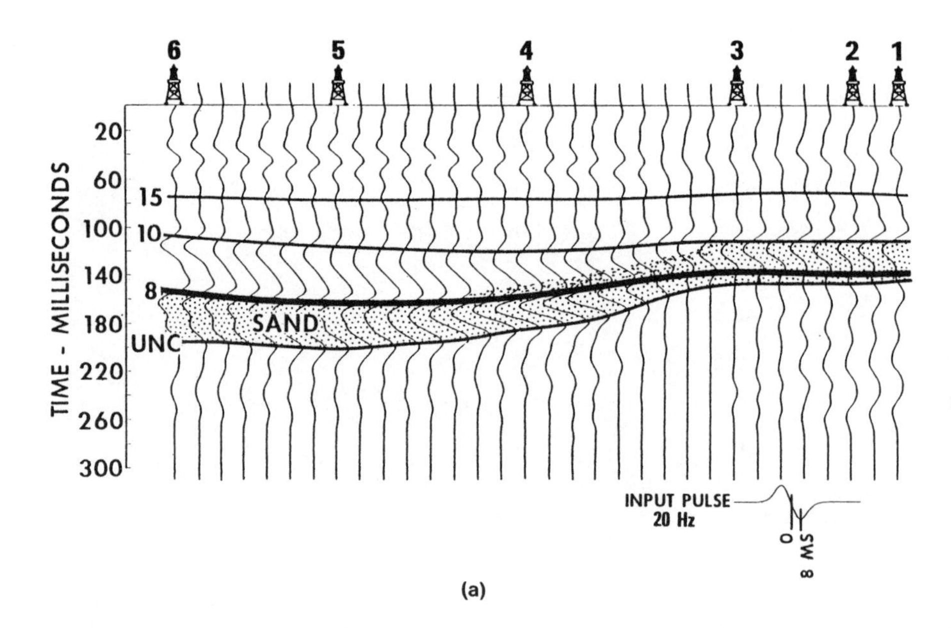

Figure 3.8. Synthetic seismic section simulating line shown in figure 3.7. (a) The entire line using a low-frequency wavelet. (b) Right and left portions juxtaposed. A jump correlation between wells 5 and 1 would likely follow a facies line because of the high amplitudes at the top of the basal sand and at the unconformity. (After Vail, Todd, and Sangree, 1977; reprinted by permission of The American Association of Petroleum Geologists)

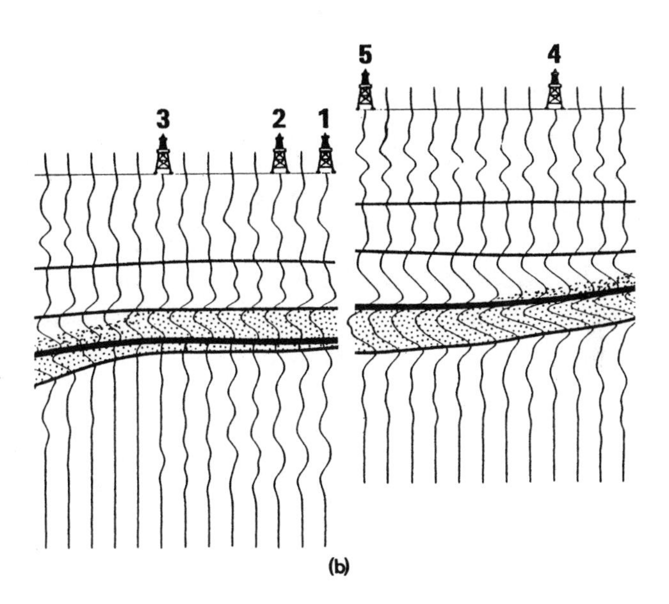

Figure 3.8 (continued).

Figure 3.8a is a model of the situation shown in figure 3.6, intended to simulate figure 3.7. If our seismic sampling were much coarser, for example, of the degree of coarseness often involved in well correlations, we might see a different pattern (figure 3.8b). The reflection from the top of the basal sand is stronger than most of the other reflections, and we might correlate this strong reflection, thereby correlating along the facies line. However, since we have the intervening traces, we pick reflections paralleling the time surfaces.

Facies boundaries are often very irregular or gradational and hence do not produce coherent reflections over appreciable distance. However, when a facies boundary is fairly sharp and widespread, it does, of course, produce a reflection.

Figure 3.9 shows a seismic section in which a reflection (II) follows a progradational pattern rather than being parallel to overlying and underlying reflections (I and II). Well control indicates that reflection II cuts across a number of facies units (figure 3.10) and this reflection parallels a time surface.

Picking Seismic Sequences

Presumably, conditions were relatively constant during the laying down of a depositional unit but then changed before the next depositional unit was deposited. Since we identify depositional units with seismic sequences, we expect a seismic sequence to gradually onlap previous sediments in the landward direction, have a thickest portion, and thin again in the seaward direction as the amount of sediment decreases because of remoteness from the sediment source. We expect onlap, erosional effects, and so on, at the boundaries between seismic sequences. The angular pattern of reflections at sequence boundaries (figure 3.11) is the key to isolating se-

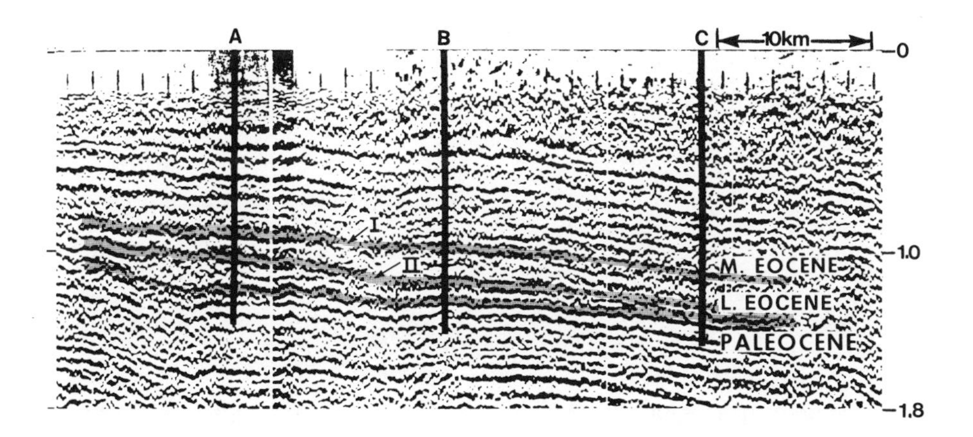

Figure 3.9. Seismic section connecting three wells. (From Vail, Todd, and Sangree, 1977; reprinted by permission of The American Association of Petroleum Geologists)

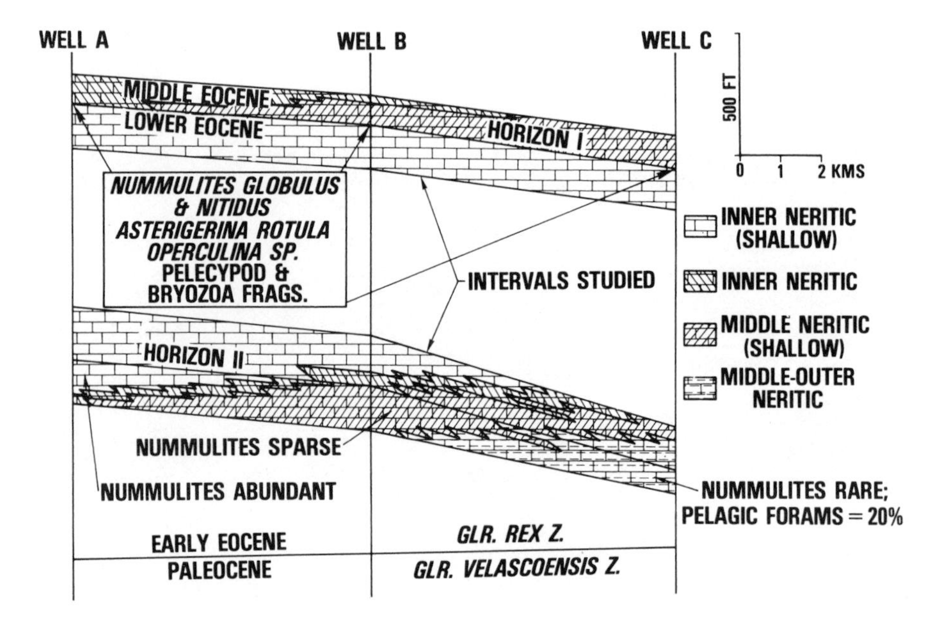

Figure 3.10. Facies distribution determined from the three wells along the seismic line shown in figure 3.9. (From Vail, Todd, and Sangree, 1977; reprinted by permission of The American Association of Petroleum Geologists)

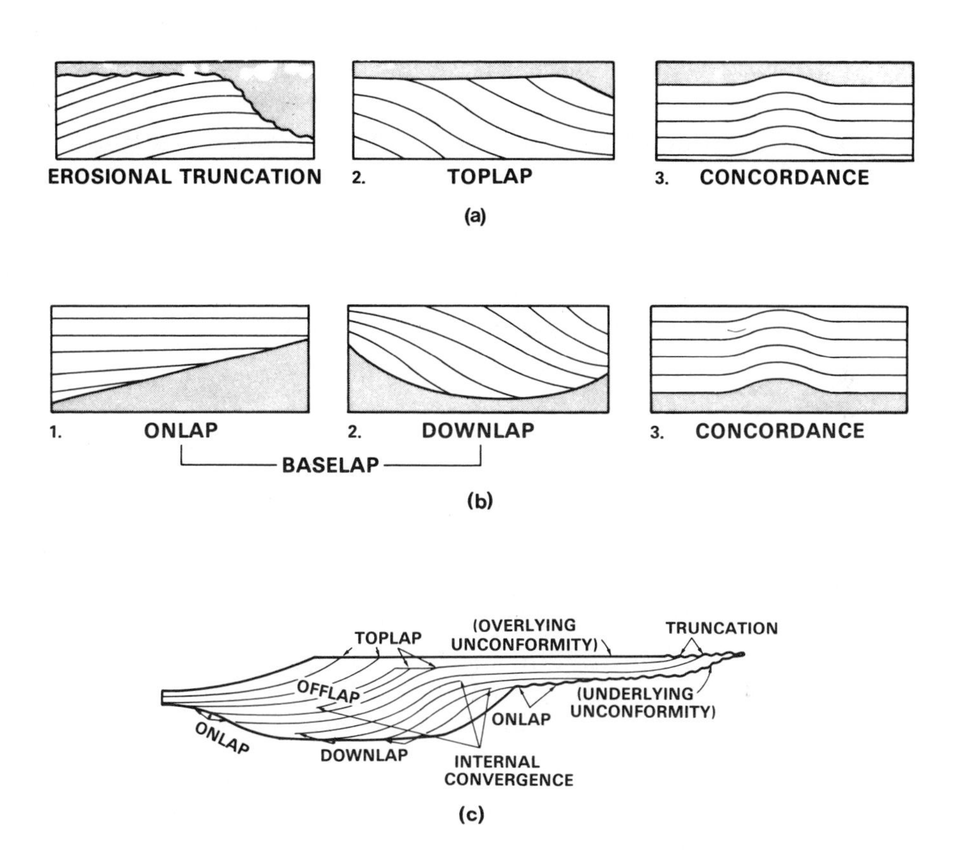

Figure 3.11. Relations of reflections within a sequence unit to the unit boundaries. (a) Relations at top of sequence unit. (b) Relations at base of sequence unit. (c) Reflection terminations within an idealized unit. Figure 5.2 shows examples of truncation, toplap, onlap, and downlap in actual seismic data. (From Mitchum, Vail, and Thompson, 1977b; reprinted by permission of The American Association of Petroleum Geologists)

quence units. Of course, the sediments within a unit may be concordant with the underlying and/or overlying boundaries in places and we may have to "push" or "phantom" through such regions to complete the bounding surfaces.

The seismic sequence analysis procedure is to pick the unconformities that bound units and thus isolate them (figure 3.12a). The seismic sequence units are often replotted as a *chronographic* (or *chronostratigraphic*) *section* (figure 3.12b), that is, as a diagram wherein age decreases upward and the lateral extent of units is displayed on the horizontal scale. Such a diagram emphasizes missing portions that have to be explained as missing because of either erosion or nondeposition.

A relative sealevel chart is also made (figure 3.12c), the relative magnitude of a rise being obtained from measurements of the coastal aggradation (figure 3.13) that can be seen on the seismic section, and the relative amount of a fall being obtained from the difference in level between a unit that is displaced a significant distance seaward compared to the top of the next older unit. The values so obtained may not indicate correctly the amount of sealevel rise or fall, but they will at least give the general picture. The relative sealevel chart should be made as objectively as possible. Adjustments for the effects of compaction, local structure, regional tilt, and so on, should be left until a more comprehensive interpretation stage is reached. Adjustments made piecemeal on individual lines are apt to be made inconsistently.

The next step is to correlate the relative sealevel chart with a eustatic level chart to age-date the sediments. The key is to correlate first the periods of major sealevel fall (figure 3.14); these are of limited number, and once they are correlated, the intervening pattern can be worked out.

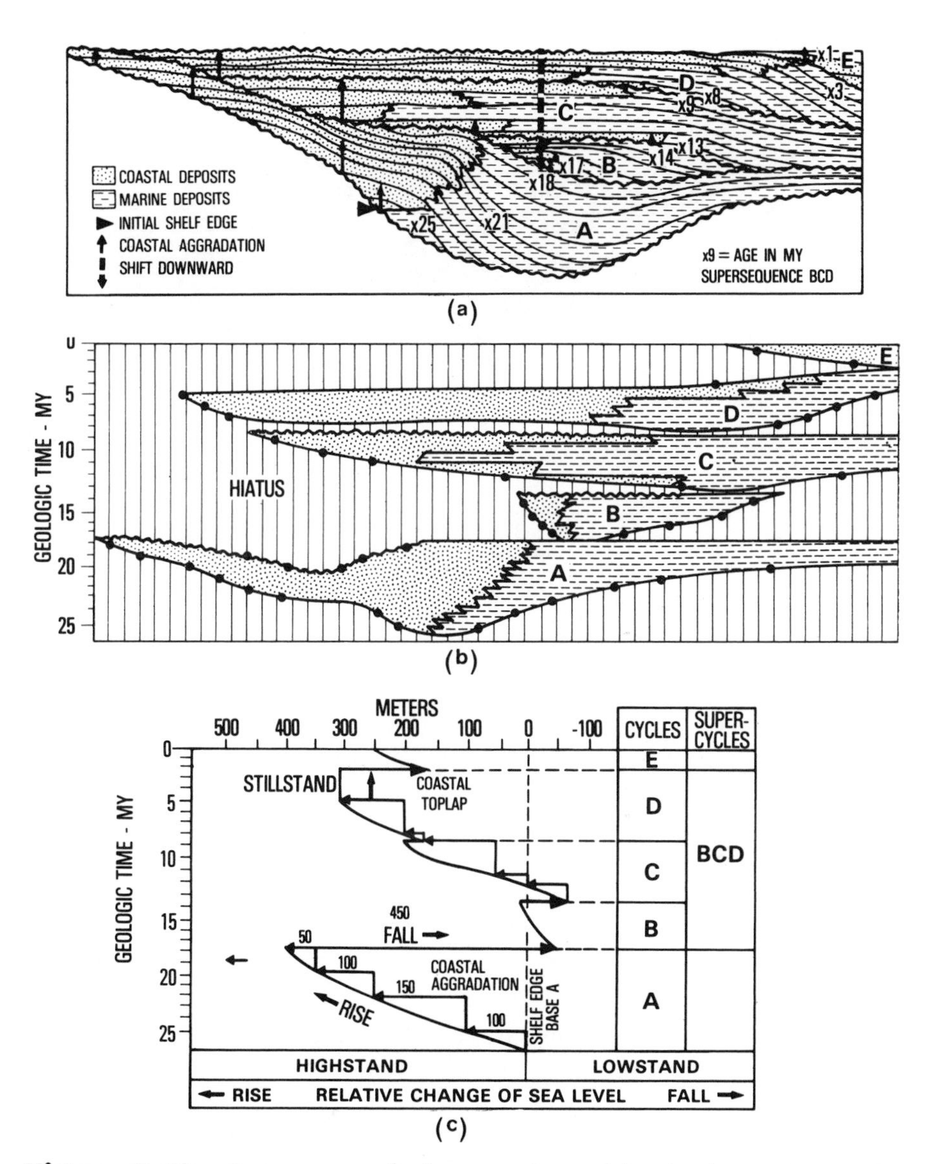

Figure 3.12. Sequence analysis procedure: (a) seismic section with sequence boundaries marked; (b) chronographic chart showing lateral extent of units (units are absent in the vertically striped region); and (c) relative sealevel chart. The time scales generally have to be added subsequently. (From Vail, Mitchum, and Thompson, 1977a; reprinted by permission of The American Association of Petroleum Geologists)

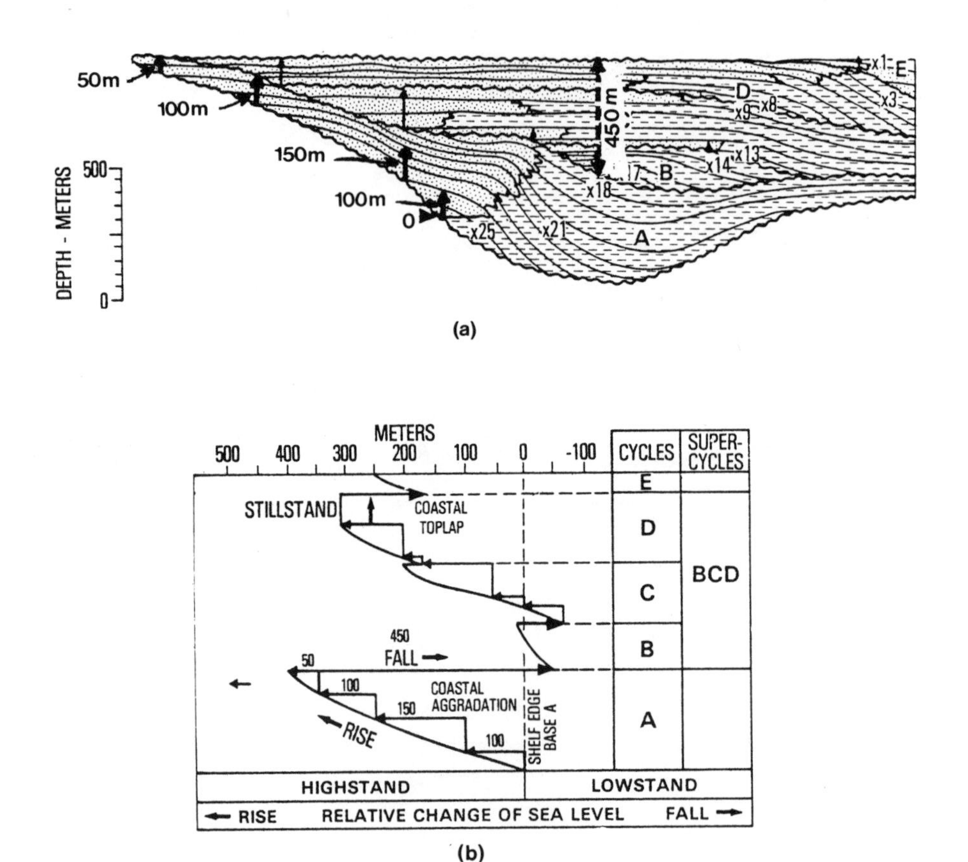

Figure 3.13. Measurements on a section to give scale to the relative sealevel chart. (a) Repeat of figure 3.12a showing measurements of relative sealevel rise during deposition of unit A and sealevel fall between end of A and beginning of unit B. Measurements are made near the onlap points to avoid erroneous thickness measurements because of regional tilting. (b) Relative sealevel chart, the time scale of which still has to be developed. (After Vail, Mitchum, and Thompson, 1977a; reprinted by permission of The American Association of Petroleum Geologists)

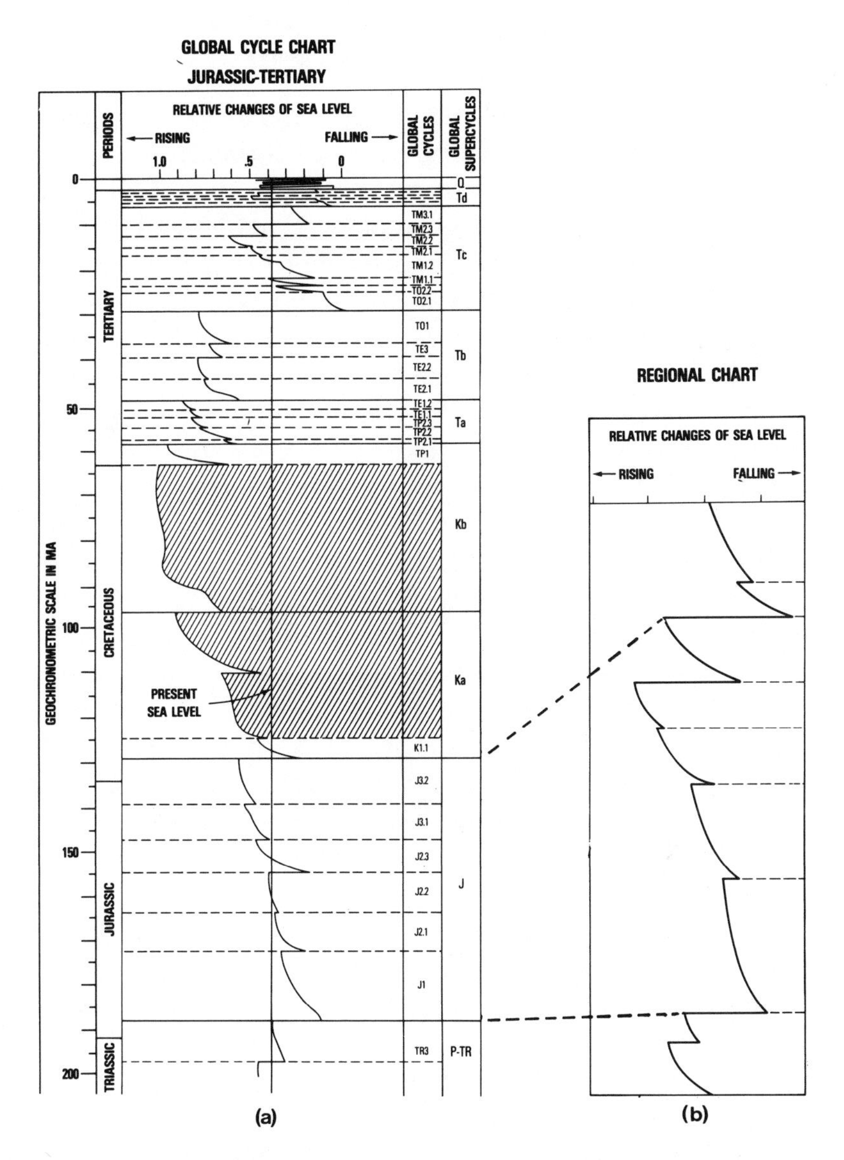

Figure 3.14. Correlation of a relative sealevel chart with master eustatic level chart: (a) portion of eustatic level chart and (b) relative sealevel chart for part of section shown in figure 4.7 with major sealevel falls correlated by dashed lines. (After Vail, Mitchum, and Thompson, 1977b; reprinted by permission of The American Association of Petroleum Geologists)

Alternate Concepts

The concept that sealevel has fallen rapidly by hundreds of meters and that this has happened periodically during geologic time is the aspect of these concepts that is most often challenged (for example, Kerr, 1980). Vail et al. visualize repeated cycles of the asymmetric pattern of rapid fall (in perhaps a few thousand years) followed by a long period of gradual rise (for a few million years), as shown in the eustatic chart (figure 3.4). They believe that these patterns are worldwide and correlate them with changes in seafloor spreading rates. As a mechanism to explain where all the water can go during a sealevel fall, they usually rely on major continental glaciation. Glaciation probably can set in with sufficient rapidity to explain the pattern asymmetry, although there is not glaciation to correlate with most of the postulated sealevel falls.

Another mechanism sometimes proposed is changes in crustal density associated with changes in heat flow and consequent isostatic adjustment. This theory can perhaps provide the space for the displaced water, but it cannot explain the pattern asymmetry. Isostatic adjustment is a slow process incompatible with the inferred rapidity of the sealevel falls.

The continent-to-continent correlations of Vail et al. are challenged by some. If, for example, the Atlantic Ocean basin can be considered a closed system, then the problems of finding a mechanism are lessened (but not eliminated). Brown and Fisher challenge the concept that a sudden seaward shift in onlap pattern evidences rapid sealevel fall. Whereas Vail et al. visualize long periods of gradual sealevel rise followed by rapid fall, Brown and Fisher visualize periods of sealevel rise interrupted by occasional stillstands of sealevel, but only rarely a fall. The two schools of thought differ especially in their concepts of the implications of marine onlap.

Brown and Fisher place more importance on submarine erosion and deposition. They do not (as Vail et al. do) interpret canyons in continental shelves and slopes as predominantly river eroded during low sealevel times, nor do they interpret sediment fans on continental slopes and in the ocean basins as being mainly river-derived sediments that have been channeled across the continental shelves through canyons during low-water times (see also figure 5.6). They assign more importance to turbidity currents and submarine gravity flow. The differences between the Vail et al. and the Brown and Fisher viewpoints is partially one of degree (although the differences are substantial). Probably both river and submarine erosion are involved in the formation of canyons across the continental shelves and in the build-up of sediment fans on the continental slopes and basin floors.

Brown and Fisher classify slopes as constructive or destructive depending mainly on the amount of sediments being brought in compared to the rate of subsidence (see figure 5.6). Where available sediments exceed subsidence effects, offlap progradational patterns result and the slopes are constructive; where the sediment supply roughly balances subsidence effects, successive layers merely stack on each other in an uplap pattern; where subsidence effects dominate, slope erosion and destructive slopes result, the evidence being marine onlap patterns. These concepts affect the evaluation of the reservoir character of deep-sea sediments (see chapter 5).

4
Seismic Sequence Analysis: The Implementation

Picking of Unconformities to Separate Seismic Sequences

Reflections generally indicate either (1) time (isochronous or stratal) surfaces or (2) unconformities. The fact that unconformities are usually good reflectors is important because the mapping of unconformities is the key to seismic sequence analysis. Unconformities represent a hiatus in deposition during which conditions are apt to have changed, so that there are different lithologies above and below, thus producing a contrast and a reflection. Furthermore, unconformities are often associated with some angularity, and the angularity of reflections provides the key for picking and identification. Because unconformities are among the best reflectors, we can often follow an unconformity on the basis of its reflection strength even where it is not marked by angularity.

The technique for mapping seismic sequences is to first recognize unconformities where angularity (onlap, erosion, and so on) is evident and then follow the unconformities through regions where there is no angularity by virtue of their amplitude standout. The procedure is to mark angularity evidences on the seismic section (figures 4.1 and 4.2) and then draw in the unconformity considering the location of these evidences and the amplitudes of events. In this way we separate different seismic sequences.

Mapping Seismic Sequences in Three Dimensions

The sequence boundaries have to be followed over a grid of intersecting lines to produce a three-dimensional map of the sequence unit. The seismic evidences may be somewhat different on lines in different directions. For example, a dip line across a fan may see progradational patterns whereas a strike line may show overlapping cusps (figure 4.3).

A map can be made for each sequence unit showing the geographical distribution, thickness (figure 4.4) and shape of the unit. The nature of the evidences that mark the sequence boundaries is also transcribed to a map (figure 4.5). The Exxon convention as used in figures 4.5 and 4.8a expresses reflection character features as a letter fraction, indicating the evidences at the upper and lower boundaries of the sequence above the line and the configuration of the reflections within the sequence below the line. This subdivides the unit into different portions on an objective basis. The different portions of the unit may be subsequently interpreted in terms of facies. For example, the portion of figure 4.5 that is labeled "concordant cycles" may be interpreted as a delta, the shape

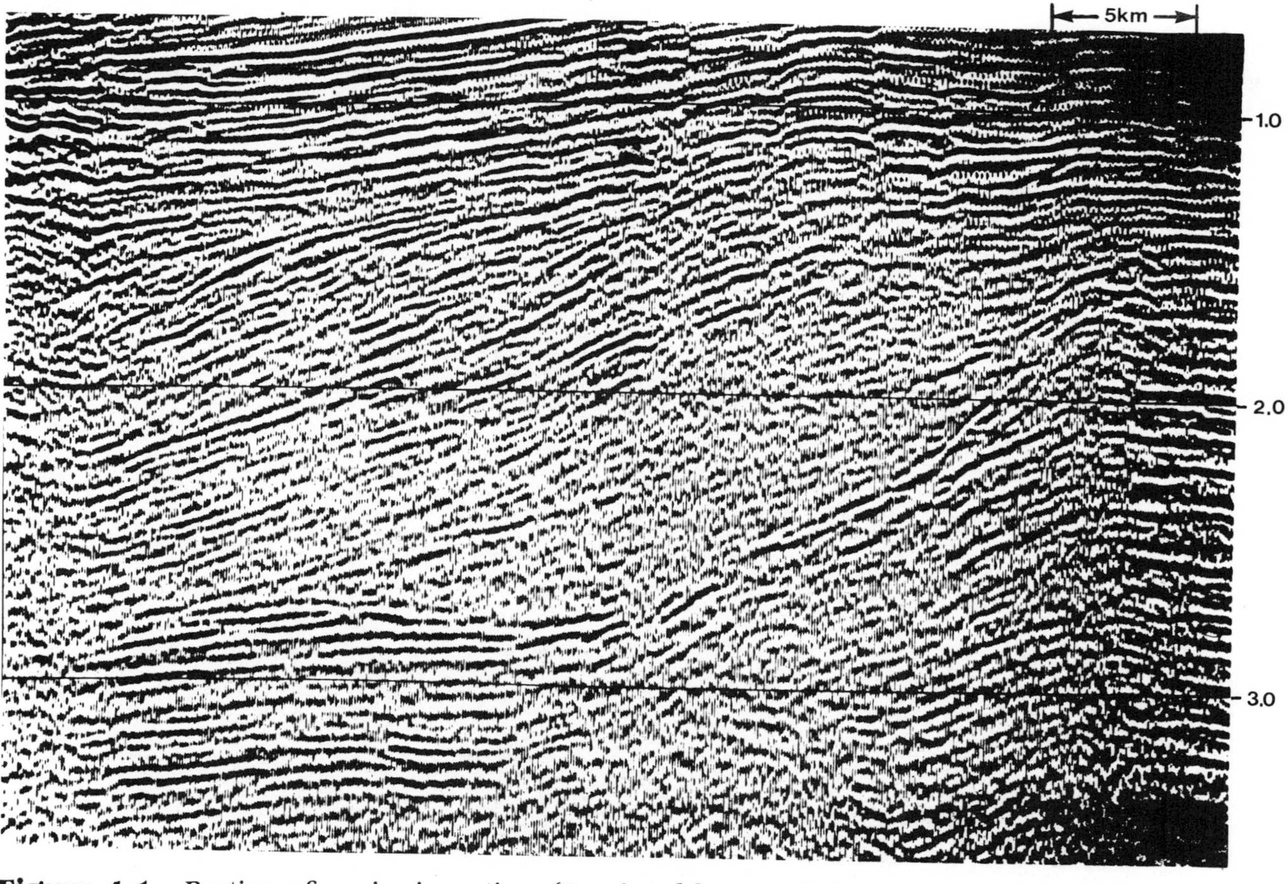

Figure 4.1. Portion of a seismic section. (Reprinted by permission of Exxon Production Research Company and The American Association of Petroleum Geologists)

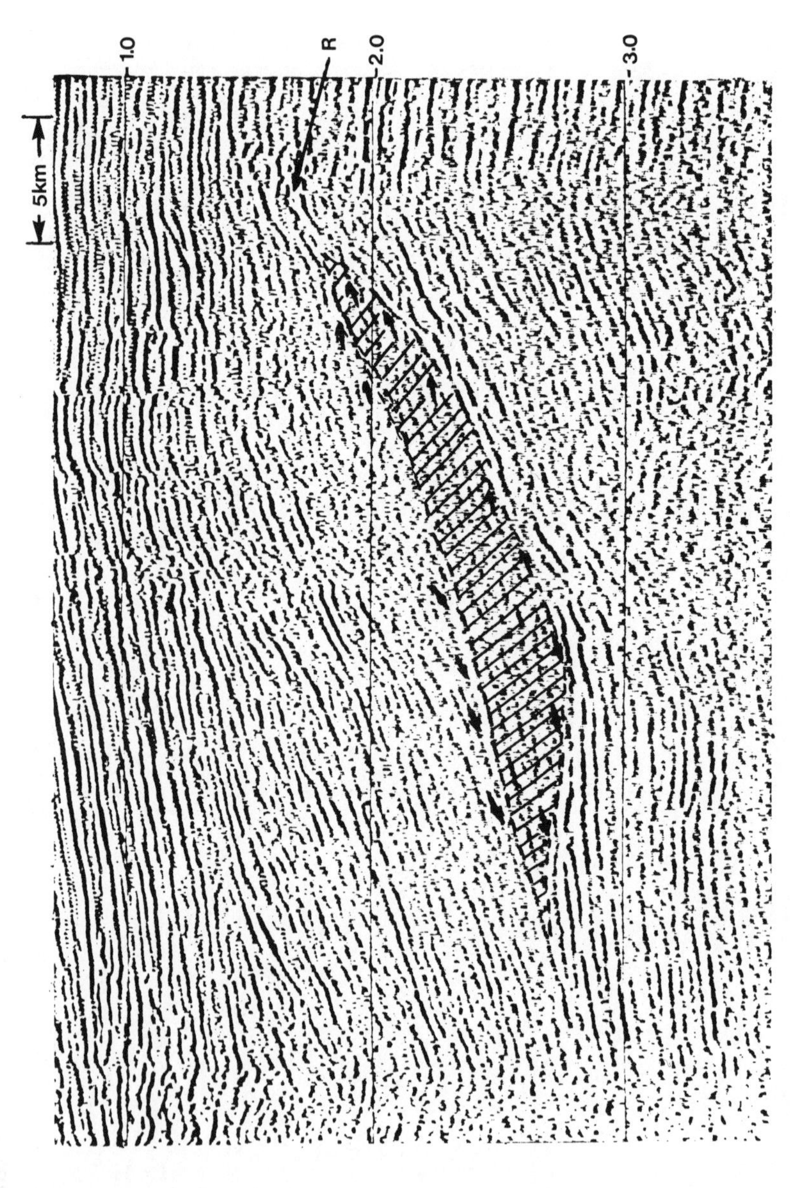

Figure 4.2. Portion of seismic section of figure 4.1 showing some reflection terminations marked by arrows to define the boundaries of one unit (shaded). This unit lies immediately seaward of a broad shelf bounded by a barrier reef (R). (Reprinted by permission of Exxon Production Research Company and The American Association of Petroleum Geologists)

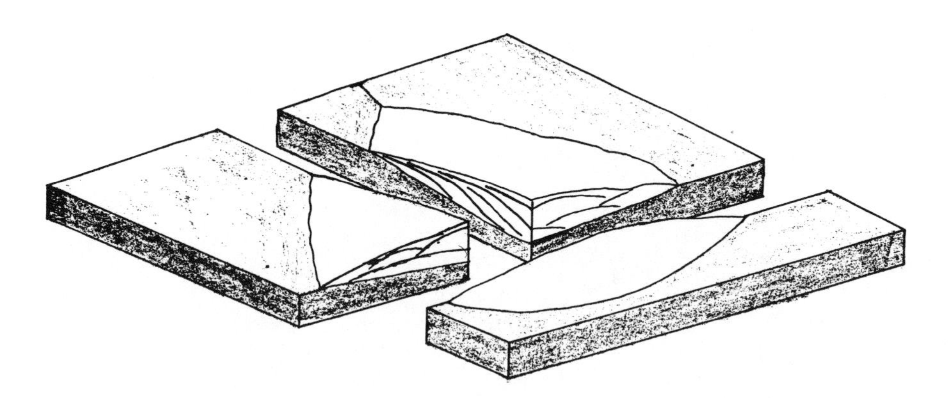

Figure 4.3. Seismic facies patterns may appear differently on lines in different directions. A dip line across a fan may show a progradational pattern; a strike line may show overlapping cusps.

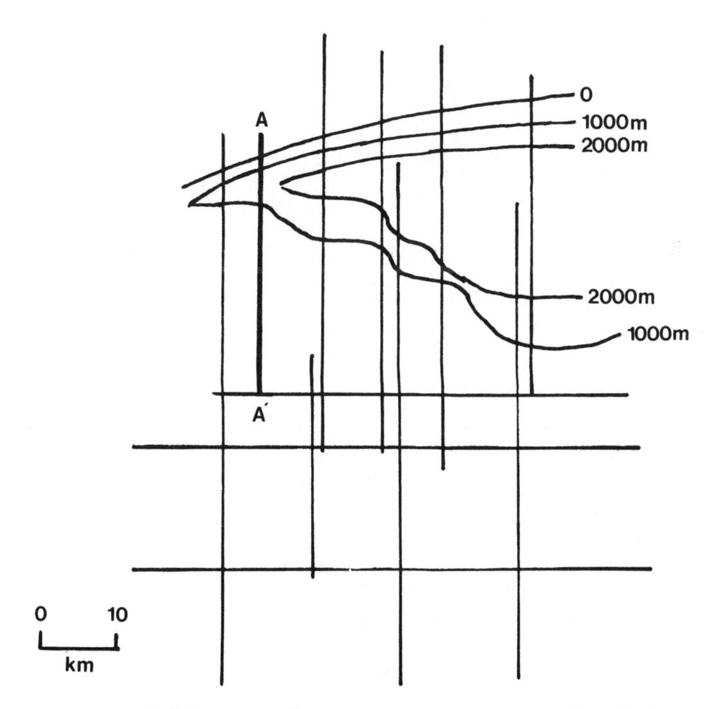

Figure 4.4. Thickness of Lower Cretaceous Valanginian-Aptian sequence. Line AA′ is that shown in figure 4.1.

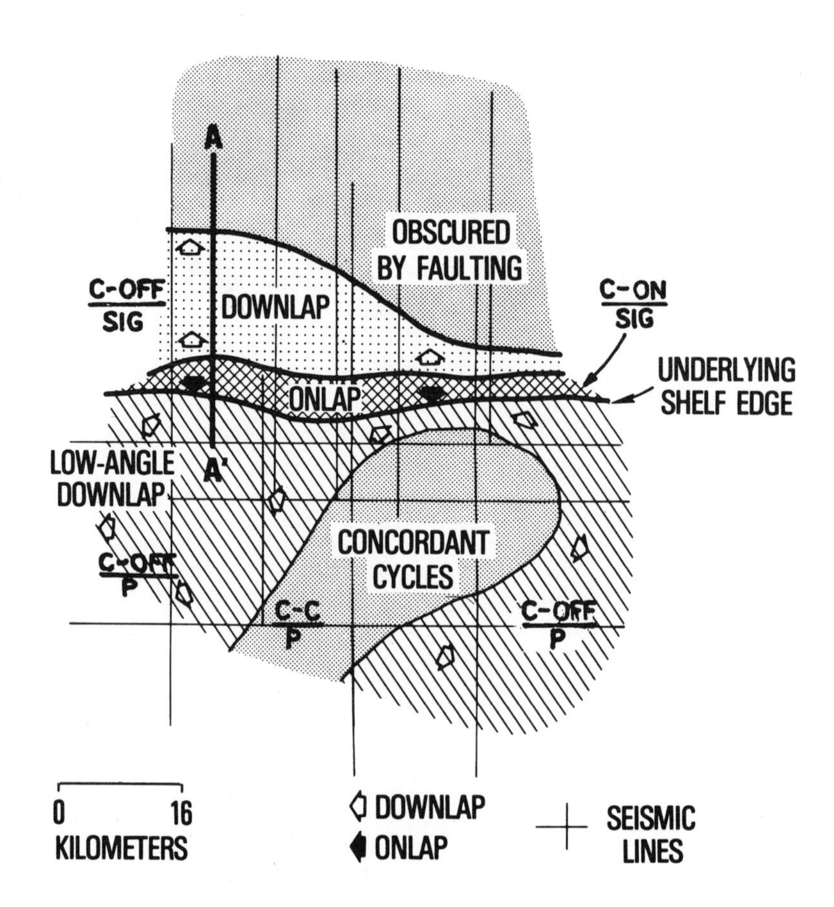

Figure 4.5. Map of a seismic sequence showing the evidences for the unit. In the fractional expressions, the two terms above the line signify the evidences at the upper and lower boundaries (onlap, offlap, downlap, toplap, erosion, and concordance) and the term below the line signifies the reflection configuration within the unit (parallel, divergent, oblique, and sigmoidal). These terms are discussed in chapter 5. Line AA′ is shown in part in figure 4.1. (After Mitchum and Vail, 1977; reprinted by permission of The American Association of Petroleum Geologists)

of the unit and its relation to other nearby units providing assistance in this identification. Other available information, such as well control, seismic velocity measurements, and so on, is incorporated, and thus a three-dimensional picture of each time-stratigraphic unit is built. The interpretation as to depositional setting, facies, and rock type is based upon ordinary geologic reasoning once the picture of the unit is developed, taking into account the relationship of the unit to neighboring units.

Examples

Figure 4.6 shows a seismic line near the portion of the line shown in figure 4.1. This line is also shown in figure 4.7, subdivided into seismic sequence units with age assignments added. The age assignments are constrained by (1) the free surface, (2) age determinations from outcrop or well information, and (3) the fact that the number of choices is limited as to the units following the major downward sealevel shifts (characterized by abrupt major seaward shifts in the depositional locales). A chronographic chart for this line is shown in figure 4.8. A relative sealevel chart for part of the section is shown in figure 3.14b, correlated with the master eustatic chart of figure 3.14a. Generalized facies assignments can be made (for example, to classify units J3.1 and J3.2 as shelf facies with reef growth near their left edge, or to classify the portion of K1.2 left of the four growth faults as slope facies with possible sands in their upper portions). More detailed facies analysis is often possible (see chapter 5). Further information can be added by velocity measurements.

Figure 4.9 shows a section in the North Sea. Event B is an unconformity evidenced by both truncation of sediments below it and onlap of sediments above it. The onlap might be either regional or due to a local positive feature; frequently one cannot tell by looking at only a small portion of section,

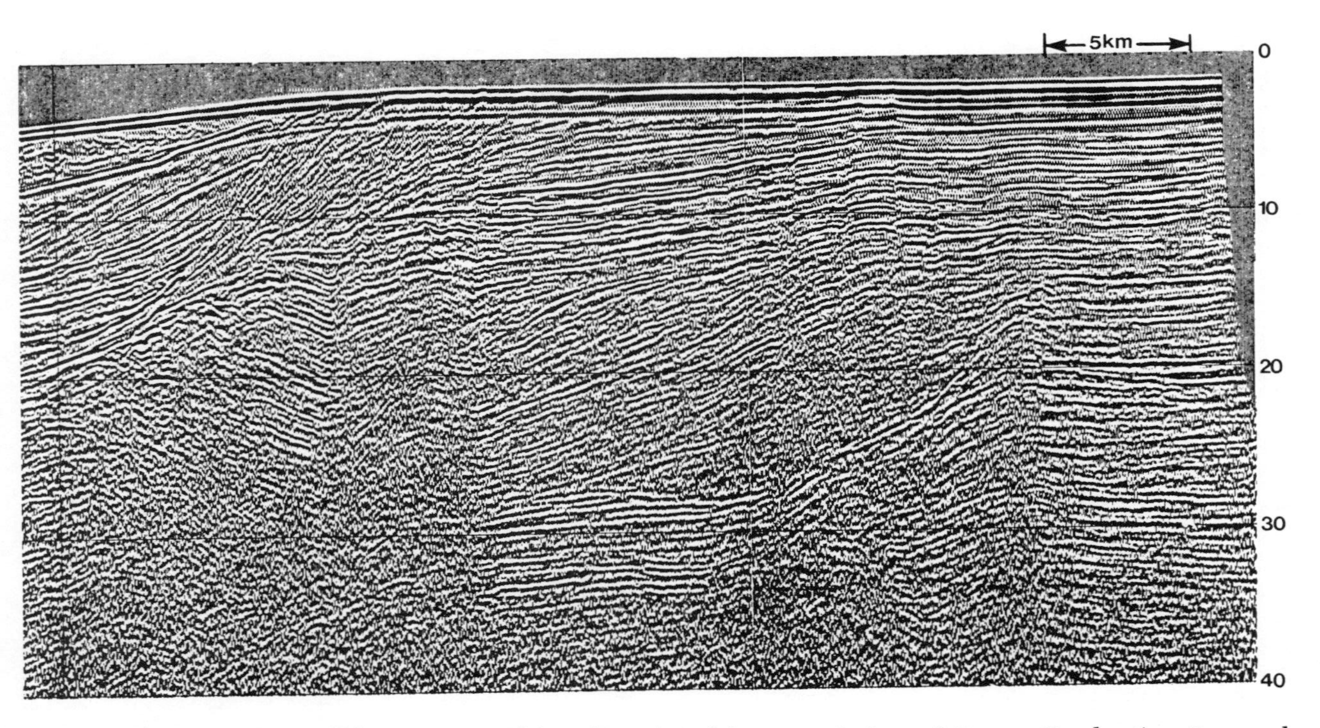

Figure 4.6. Section offshore West Africa. (Reprinted by permission of Exxon Production Research Company and The American Association of Petroleum Geologists)

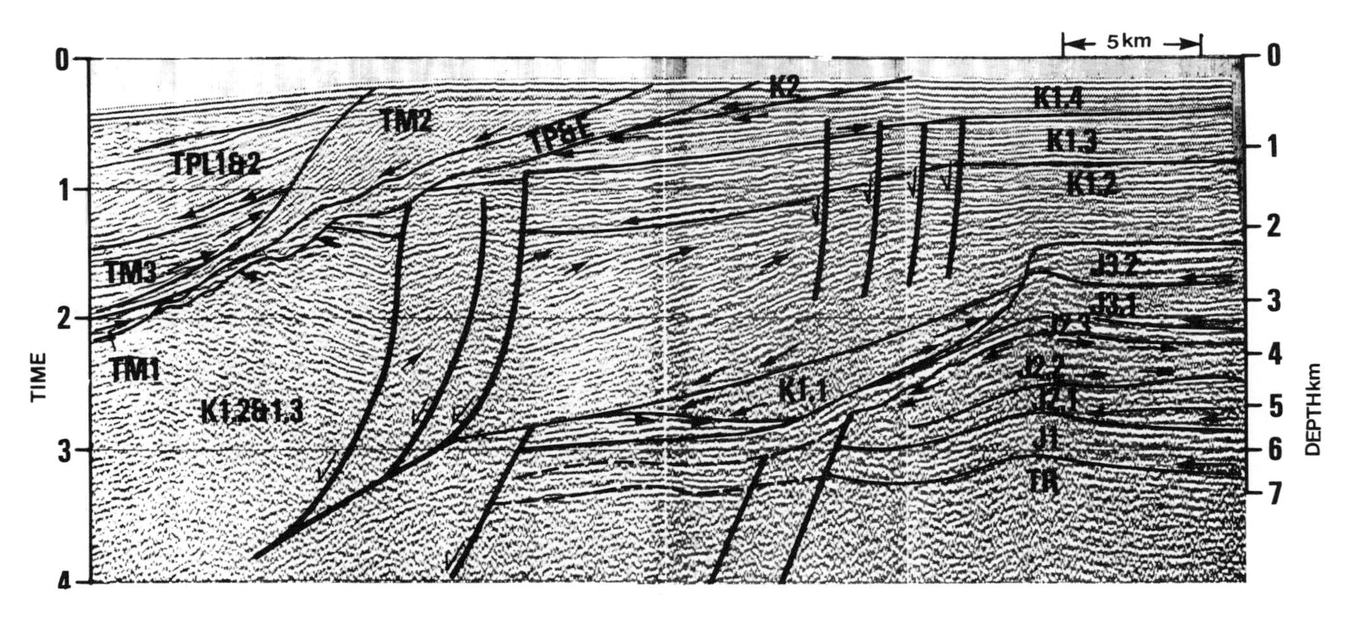

Figure 4.7. Section where angularities between seismic reflections (indicated by arrows) allow picking of unconformities that separate different depositional units. The letters indicate the ages of the units: TR, Triassic; J, Jurassic; K, Cretaceous; TP&E, Tertiary Paleocene and Eocene; TM, Tertiary Miocene; and TP, Tertiary Pliocene. (From Mitchum, Vail, and Thompson, 1977; reprinted by permission of The American Association of Petroleum Geologists)

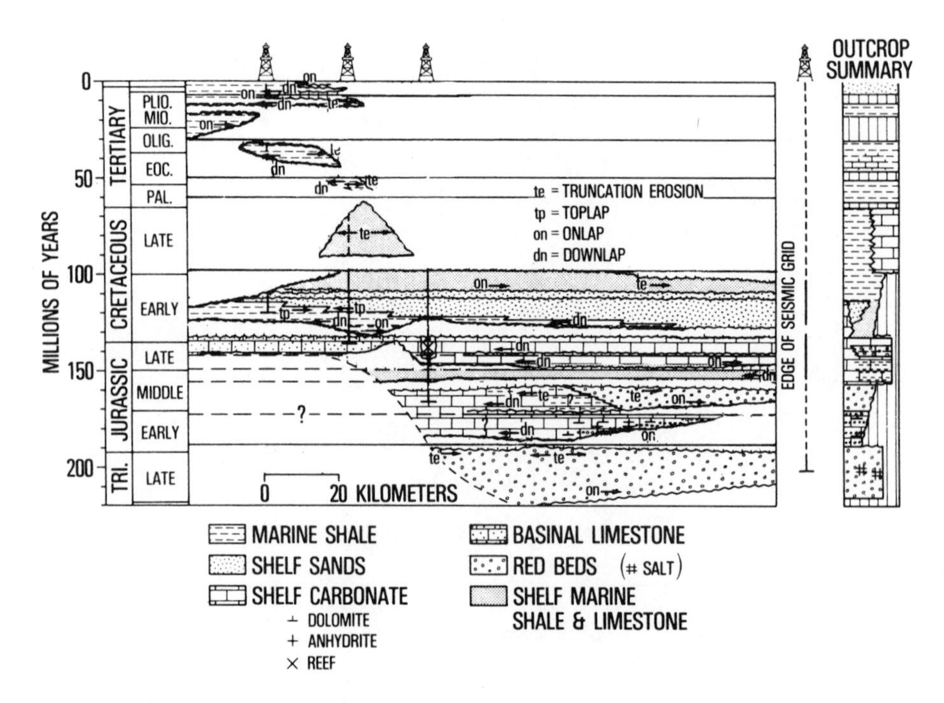

Figure 4.8. Chronographic chart of the section shown in figure 4.7. (After Mitchum and Vail, 1977; reprinted by permission of The American Association of Petroleum Geologists)

but with enough data the distinction is usually clear. In this case the onlap is more-or-less symmetrical on both sides of the anticline, suggesting that the anticline was a positive feature when these sediments were being deposited. Deposition around such a positive feature might have involved enough depositional energy to have provided some sorting of the coarser-grained sediments, assuming that coarse sediments were available, and thus these might be reservoir-quality sediments (although here they are too shallow to be of interest).

Between unconformities D and E in figure 4.9, the pattern is one of thinning from right to left in downlap fashion. These units thin in the same direction over a wide region without regard to local structure. This implies a regional pattern involving the seaward edge of deposition, with the thinning caused mainly by starvation because of remoteness from the source. These are probably fine-grained sediments that lack permeability and are not of reservoir quality.

Figure 4.10 shows the Woodbine depositional unit in East Texas; sequence boundaries are evidenced by angularity at various places. The nature of the boundary evidences from this and other lines in the area is plotted in map form (figure 4.11a); this separates the various portions of the unit on the basis of measureable seismic characteristics. Finally these subunits are given interpretational significance (figure 4.11b) based on these evidences and the shape and position of the various portions relative to each other.

Summary of Procedure

Seismic sequence analysis procedure can be summarized in the following steps:

(1) Recognize unconformities based on their angularity and regard these as unit boundaries.

(2) Extrapolate these boundaries where the reflections are conformable, so as to define the sequence units completely.

(3) Characterize portions of the units by evidences at the upper and lower boundaries and by seismic facies characteristics within the units (see chapter 5).

(4) Map the units so as to see their shapes, orientations, and so on.

(5) See how the units relate to neighboring units, known geological information, velocity information, and so on.

(6) Finally, synthesize these evidences into an interpretation based on stratigraphic concepts.

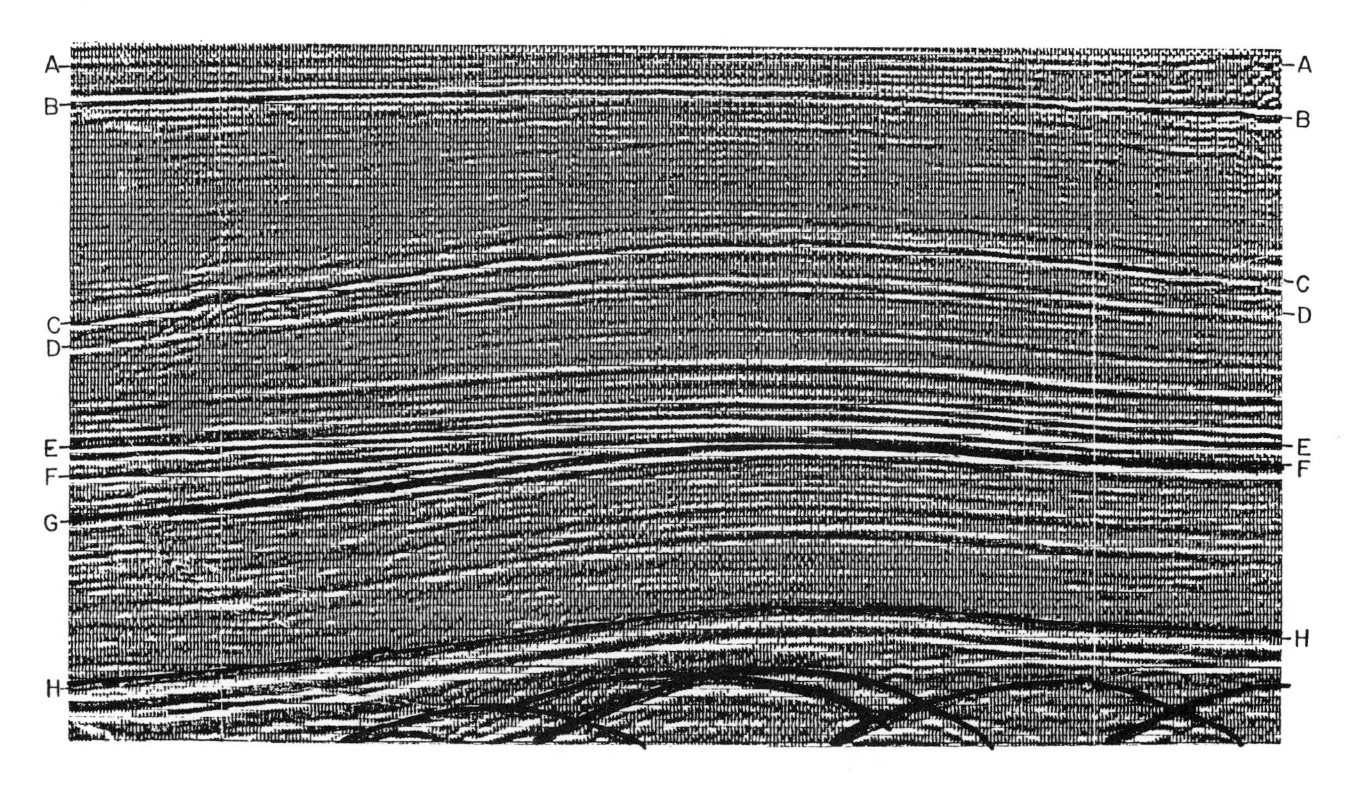

Figure 4.9. Seismic line in the North Sea. The letters indicate sequence boundaries. The diffractions at the bottom of the section indicate salt flowage. (Reprinted by permission of Seiscom Delta)

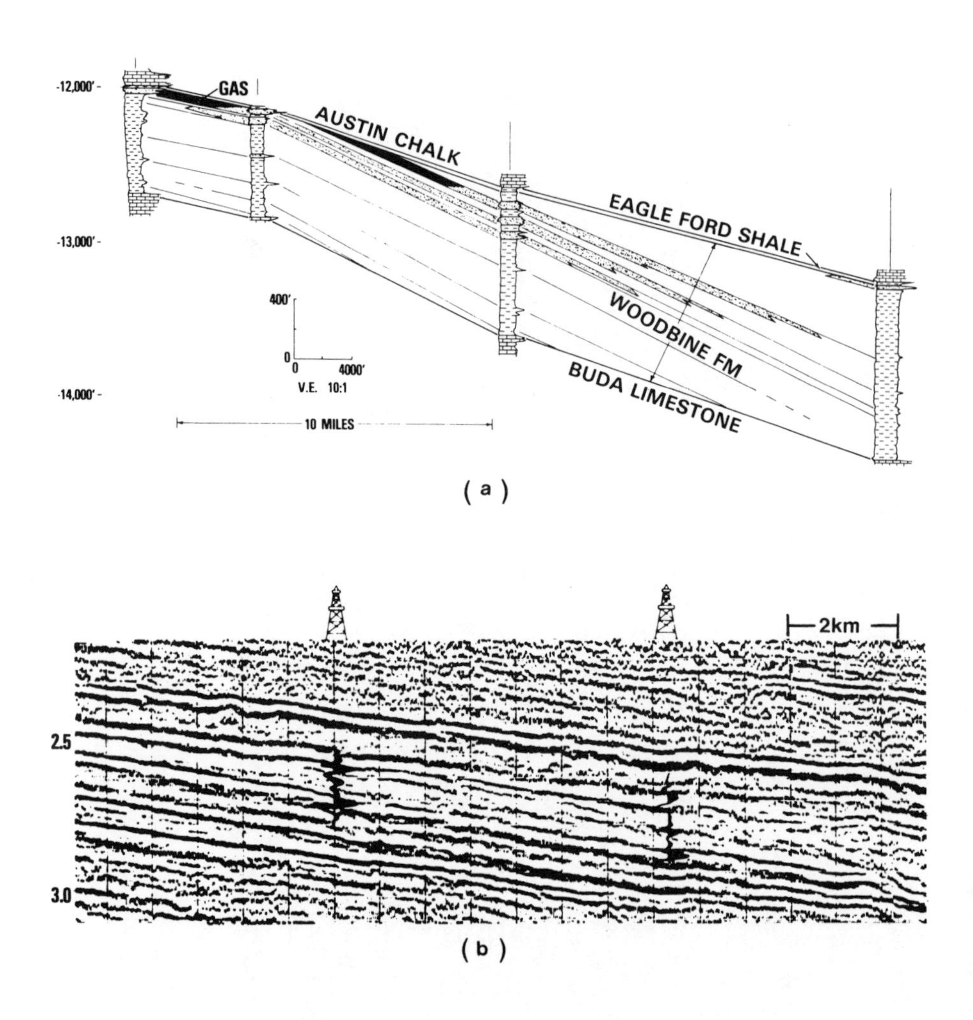

Figure 4.10. Section in East Texas. (a) Section through four wells. (b) Portion of a seismic line through the two right-hand wells shown in (a). Portions of synthetic seismograms made from logs in these wells are superimposed on the section. (From Ramsayer, 1979; reprinted by permission of The Offshore Technology Conference)

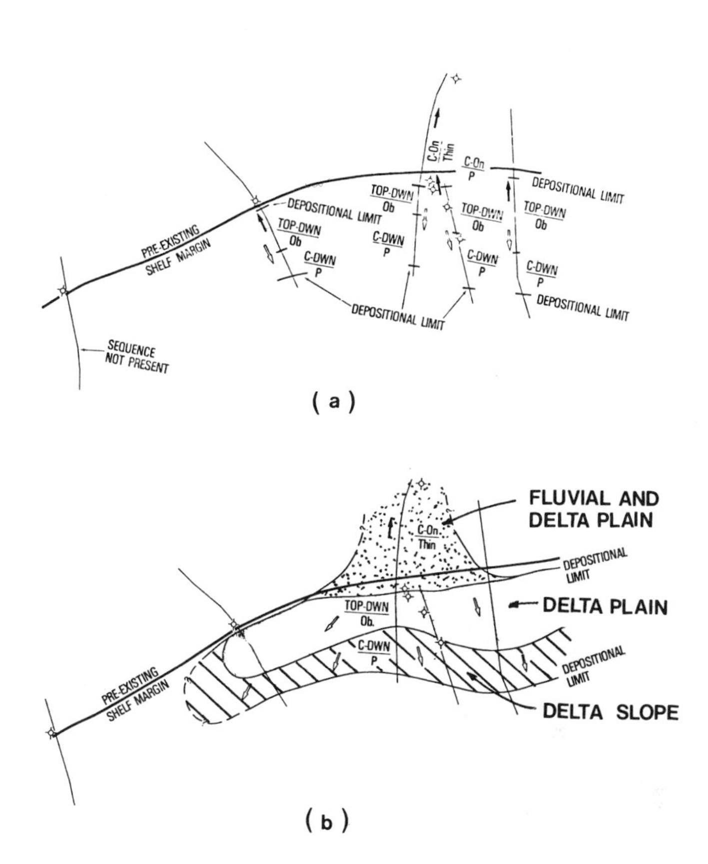

Figure 4.11. Map of Woodbine unit in East Texas based on seismic evidences. (a) Evidences along five seismic lines posted on map. TOP = toplap, C = concordance at top of unit, On = onlap at base of unit, DWN = downlap at unit base, Thin = unit is not thick enough to see internal patterns, Ob = oblique internal pattern, and P = parallel internal reflection pattern. Solid arrows indicate direction of onlap, open arrows indicate direction of downlap. (b) Facies interpretation of the seismic evidences. (After Ramsayer, 1979; reprinted by permission of The Offshore Technology Conference)

5
Seismic Facies Analysis and Reef Patterns

Seismic Facies

The AGI *Glossary of Geology* (1972) defines *facies* as:

> The sum of all characteristics exhibited by a sedimentary rock and from which its origin and environment of formation may be inferred; the general aspect, nature or appearance of a sedimentary rock produced under or affected by similar conditions; a distinctive group of characteristics that differs from other groups within a stratigraphic unit.

Seismic facies concerns those distinctive characteristics that can be seen in seismic data.

A *seismic facies unit* is a mappable three-dimensional unit of reflections whose characteristics differ from that of the adjacent facies. We use seismic facies distinctions as well as unconformities to separate seismic sequence units. (The terms "seismic sequence" and "seismic facies unit" are sometimes used for each other, but here we try to distinguish between them; a *seismic sequence unit* is a package of reflections that results from the sediments within a time-stratigraphic depositional unit, that is, they represent some certain (though possibly unknown) age interval, whereas *seismic facies* describes the seismic characteristics that result from the depositional environment, for example, marine or non-marine.)

The objective of seismic facies analysis is regional stratigraphic interpretation, specifically, defining the depositional environment, the lithology, and the geologic history.

Types of Reflection Characteristics

Reflections can be characterized by amplitude, continuity, dominant frequency, abundance, smoothness, and so on. These distinctions, especially reflection amplitude and continuity, are particularly useful for determining the depositional environment of sediments laid down on shelves, where they form the basis of the classification scheme to be discussed subsequently. Reflection configuration provides another means of characterizing reflections that is particularly useful for sediments laid down on shelf margins and slopes. The overall shapes of units and the relation of their shape to internal reflection configuration is particularly useful for classifying sediments laid down on basin slopes and floors. Before discussing the classification schemes, we classify reflection patterns in the manner of Mitchum, Vail, Sangree, and Widmier, dividing layered patterns according to their configuration both at sequence boundaries and within a sequence, and we classify the overall three-dimensional shape of sequences.

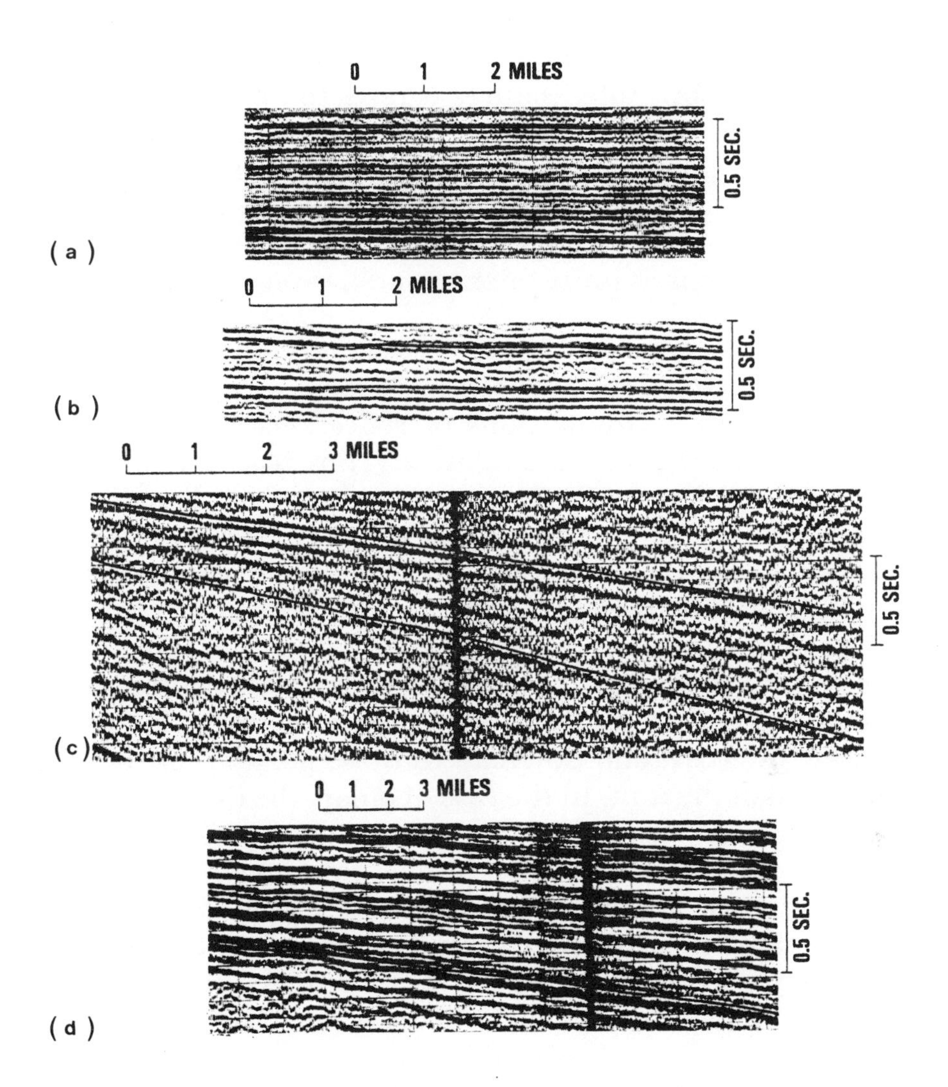

Figure 5.1. Reflection configurations within sequence units: (a) and (b) parallel and subparallel reflections and (c) and (d) divergent reflections. (After Mitchum, Vail, and Sangree, 1977; reprinted by permission of The American Association of Petroleum Geologists)

Simple Reflection Configurations

The most common types of reflection configurations consist of parallel, subparallel (figures 5.1a, 5.1b), or divergent reflections. Divergent reflections (figures 5.1c, 5.1d) gradually spread out, almost always in the down-dip direction, indicating gradual basin subsidence during deposition, with gradual basinward tilting. New reflections appear between other reflections as units thicken and become resolvable (see chapter 6).

Configurations at Unit Boundaries

The bounding configurations of reflections at the base of seismic sequence units (see figure 3.11) are base onlap, downlap, or parallelism to the boundary (concordance). Onlap examples are shown in figures 5.2a and 5.2b and downlap examples in figures 5.2c and 5.2d. The distinction between onlap and downlap is based on the geometry, that is, whether the onlapping bed at its termination is flat to dipping upward (onlap) or is dipping downward (downlap). Sometimes onlap relates to those portions of the sequence nearest the coast and downlap to those portions at the seaward end of the unit. In the case of onlap, the distinction implies thinning because there was not much room for deposition. In the case of downlap it implies thinning because there was not much sediment available. In the early phases of an interpretation, distinction should be made on the basis of geometry only, without regard for the implied reason. Often no distinction is made; then both onlap and downlap are called baselap.

The bounding configuration of reflections at the top of seismic sequence units are toplap, erosional truncation, or parallelism to the boundary. The distinction between erosional truncation (figures 5.2e, 5.2f) and toplap (figures 5.2g, 5.2h) is whether the sediments did or did not extend significantly higher than their present situation indicates.

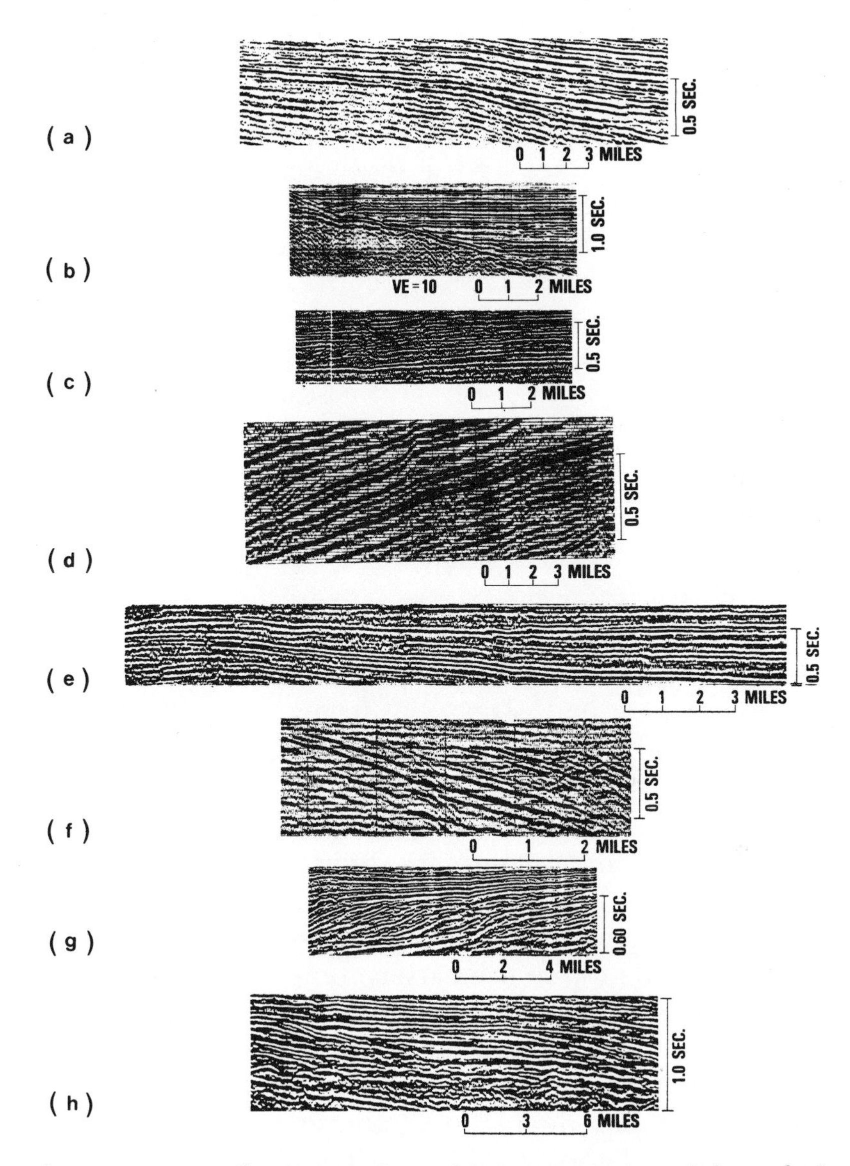

Figure 5.2. Reflection configurations at sequence unit boundaries. (a) to (d) Effects at base of unit: (a) and (b) onlap and (c) and (d) downlap. (e) to (h) Effects at top of unit: (e) and (f) erosional truncation and (g) and (h) toplap. (After Mitchum, Vail, and Sangree, 1977; reprinted by permission of The American Association of Petroleum Geologists)

Toplap implies deposition near the wave base where there was appreciable energy during or shortly after the deposition of the sediments, so that there is reasonable probability that sorting of grain sizes by the wave energy will have occurred. Erosional truncation, on the other hand, implies that the sedimentary unit formerly extended significantly beyond its present limits—possibly being exposed above the water—and that the top of the unit was removed. The distinction is not always clear. Resolution (chapter 6) usually limits the detail that can be seen at the boundary. Ringing of an unconformity reflection because of inadequate processing also sometimes obscures what happens. (Ringing gives a wavelet a duration of several cycles.)

Complex Reflection Configurations

Patterns within a sequence unit are occasionally very complex (figure 5.3) and are given a variety of descriptive terms such as hummocky, shingled, contorted, wavy, lenticular, disrupted, chaotic, irregular, variable, reflection-free, and so on.

The most common complex internal patterns indicate progradation (outbuilding). These are divided into two classes, oblique and sigmoid, and are illustrated in figures 5.4 and 5.5. The sigmoid pattern is distinguished by a very gentle "S" shape, with the top of the pattern tending to parallel the sequence boundary, whereas the oblique pattern shows toplap or angularity with the upper boundary of the sequence. The sigmoid pattern represents deposition in quiet (usually deep) water with low depositional energy so that the top of the unit was not disturbed during deposition; the sigmoid pattern therefore is indicative of poor sorting and fine-grain sediments. The oblique pattern, on the other hand, represents deposition near the wave base in a high-energy environment, and thus sorting of grain sizes may have occurred

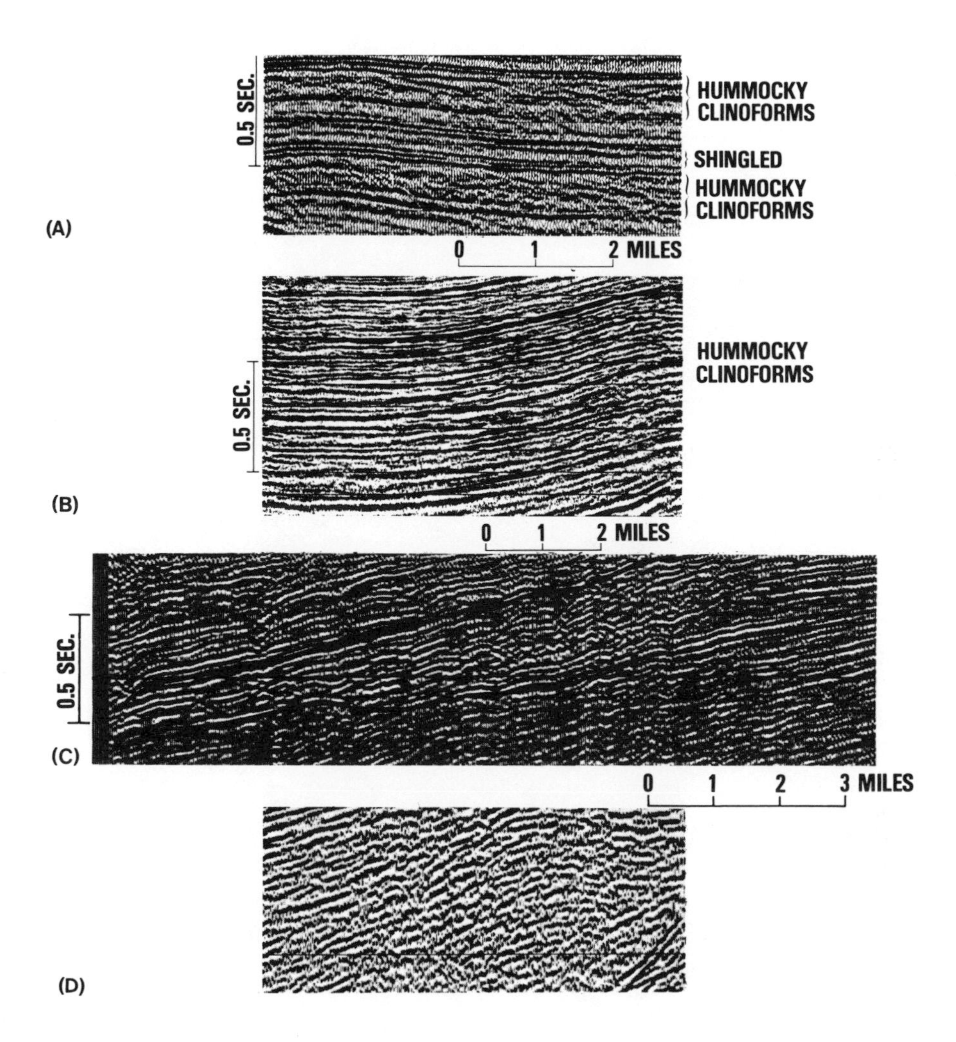

Figure 5.3. Complex reflection configurations within sequence units: (a) and (b) hummocky clinoforms (*clinoform* implies original depositional dip, as the foreset beds of a delta complex: see glossary); (c) contorted; (d) no pattern evident. Other descriptive terms often used include wavy, lenticular, disrupted, irregular, variable, reflection-free, and so on. (After Mitchum, Vail, and Sangree, 1977; reprinted by permission of The American Association of Petroleum Geologists)

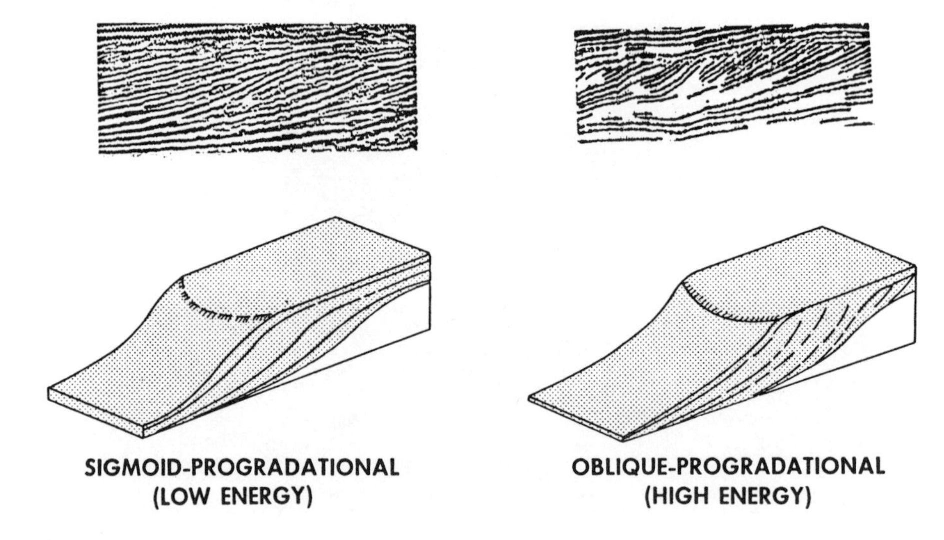

Figure 5.4. Sigmoid and oblique progradational types. The distinction is mainly based on appearances at the top of the unit. Reflections within oblique patterns are more variable than those within sigmoid patterns. (After Sangree and Widmier, 1979; reprinted by permission of The Society of Exploration Geophysicists)

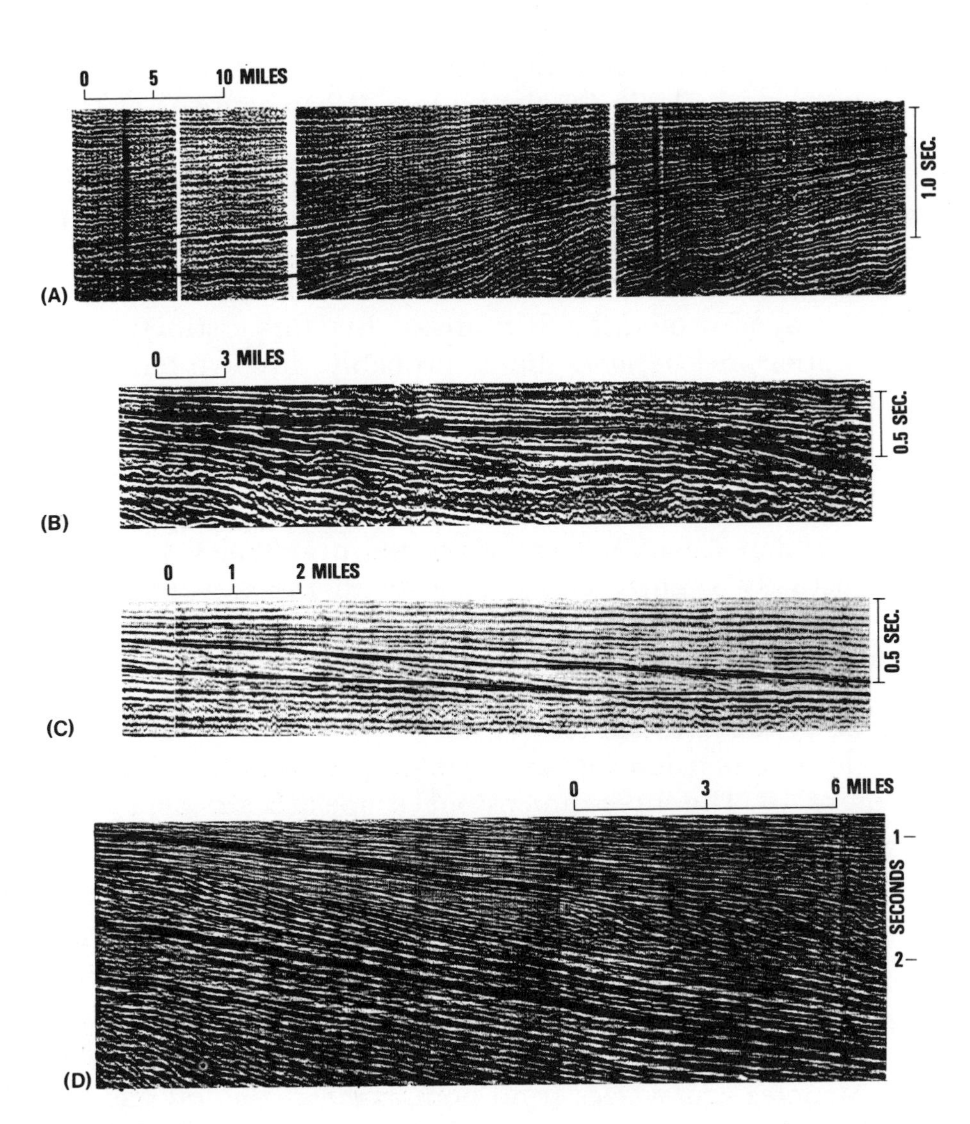

Figure 5.5. Progradational reflection configurations within sequence units: (a) sigmoid; (b) mostly oblique, some sigmoid elements; (c) mostly oblique; and (d) complex sigmoid-oblique. (After Mitchum, Vail, and Sangree, 1977; reprinted by permission of The American Association of Petroleum Geologists)

at the tops of oblique patterns. Whether oblique patterns contain fairly clean sands depends also on whether sand-size grains were available for deposition. What happens at the top of progradational patterns is not always clear because of resolution limitations (resolving power is lost as units thin); what appears to be a toplap pattern may indicate merely that the intervals are too thin for the reflections to be resolved. The distinction between oblique toplap and erosional truncation may also be difficult to make, but this distinction is less important because both probably involve enough energy to have provided sorting.

A distinction is also made between coastal onlap and marine onlap (figures 5.6a, 5.6b). Coastal onlap implies that sealevel was rising with respect to landlevel. Sediments in a coastal onlap situation were deposited near sealevel, probably actually slightly above sealevel in the case of some fluvial deposits. Marine onlap implies deposition well below sealevel.

Three-Dimensional Aspects

The three-dimensional geometry of sequence units is shown in figure 5.7. The three-dimensional shape of basin-slope and basin-fill units (figure 5.8) often provides the fundamental key to their identification (see table 5.1).

Examples

Figure 5.9 shows a portion of a long line where many wells give us subsurface information about stratigraphic change. The Menefee unit varies from nonmarine at the left end of the line through an interfingering zone of near-shoreline sands to marine at the right end of the line. The reflections in this zone vary from somewhat irregular and low/moderate amplitude in the nonmarine portion through a zone where continuity is interrupted to strong, continuous, high-

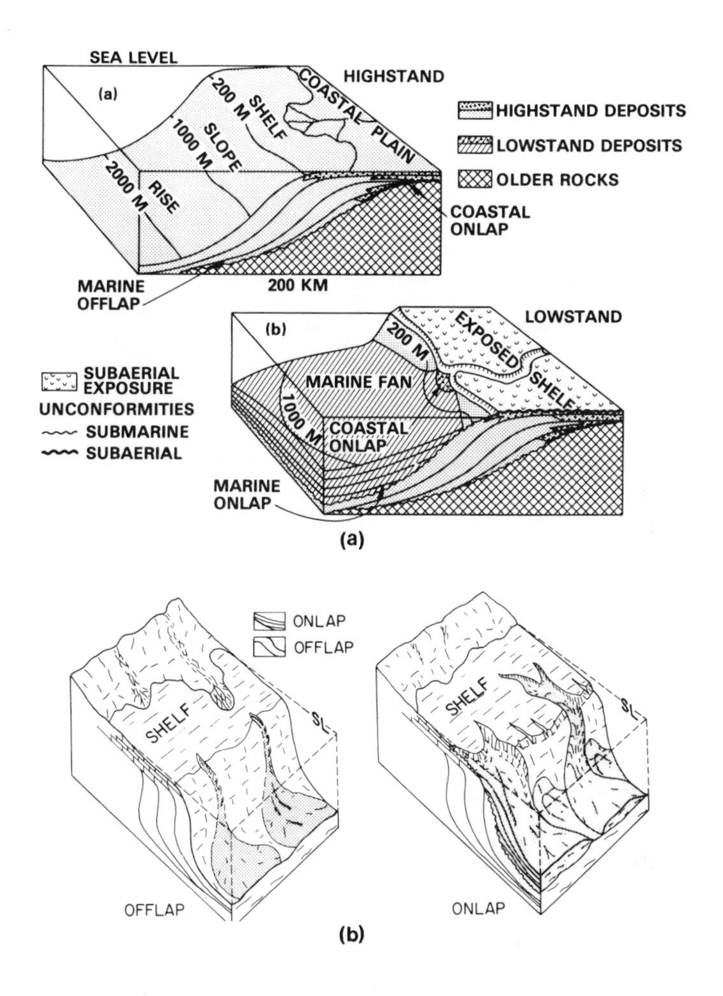

Figure 5.6. Marine offlap and marine onlap settings. (a) A highstand of sealevel produces coastal onlap and marine offlap (top) while a lowstand of sealevel results in rivers entrenching themselves across the shelves with localization of source material and the outbuilding of fans that involve marine onlap (this is the concept of Vail, et. al.). (b) Adequate sediment supply produces outbuilding of the shelf ("constructive shelf") and erosion of the shelf edge is minimal (left) while shelf outbuilding ceases when adequate sediments are no longer available and erosion of the shelf edge becomes dominant ("destructive shelf") (this is the concept of Brown and Fisher).

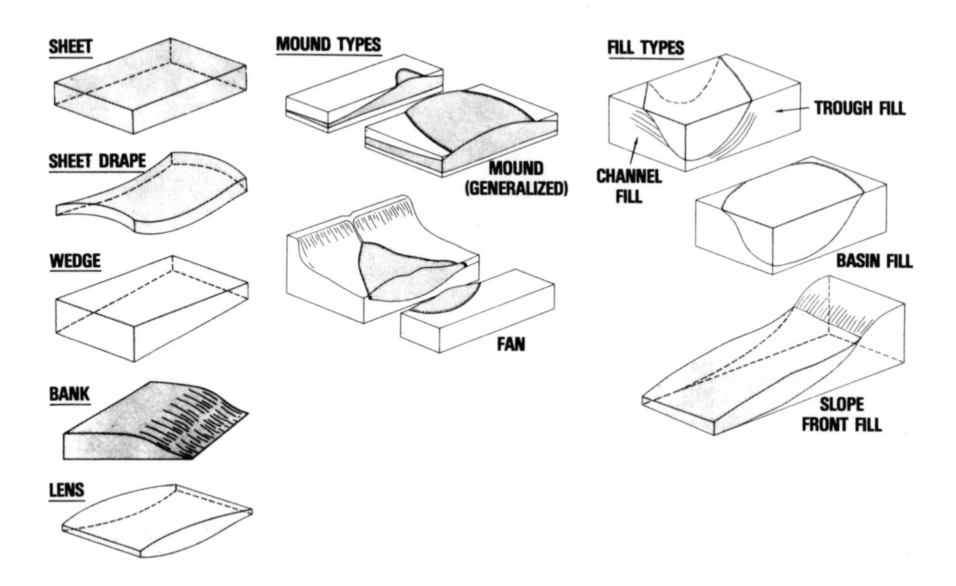

Figure 5.7. Three-dimensional geometry of seismic sequence units. (After Mitchum, Vail, and Sangree, 1977; reprinted by permission of The American Association of Petroleum Geologists)

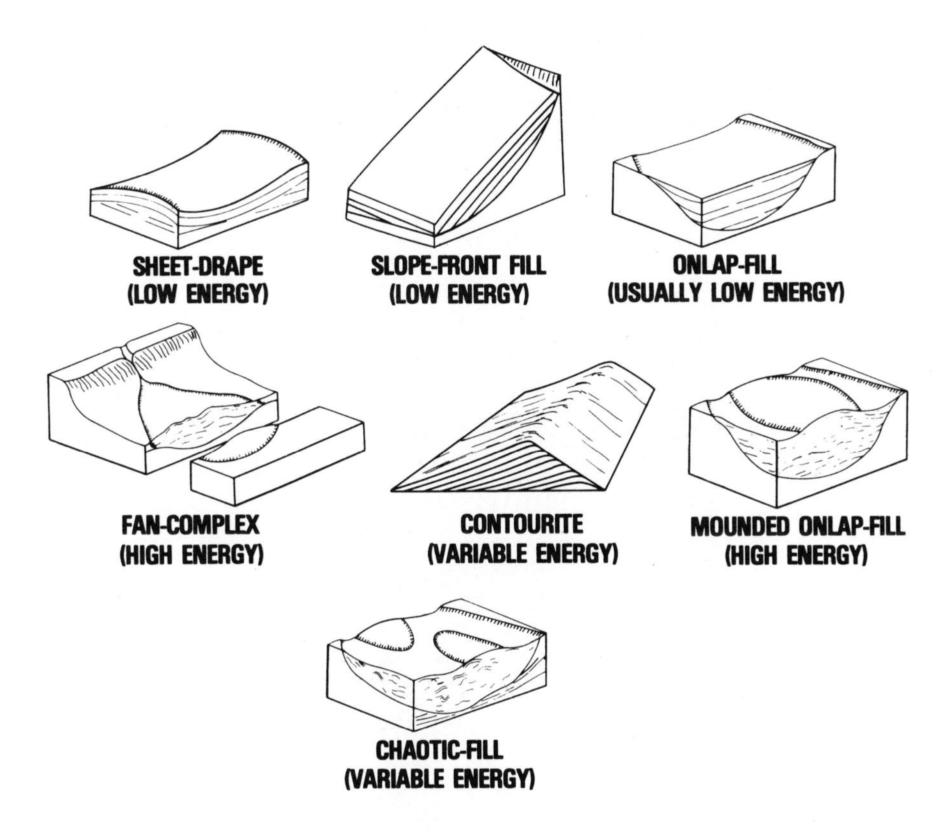

Figure 5.8. Geometry of basin slope and basin fill units. (From Sangree and Widmier, 1977; reprinted by permission of The American Association of Petroleum Geologists)

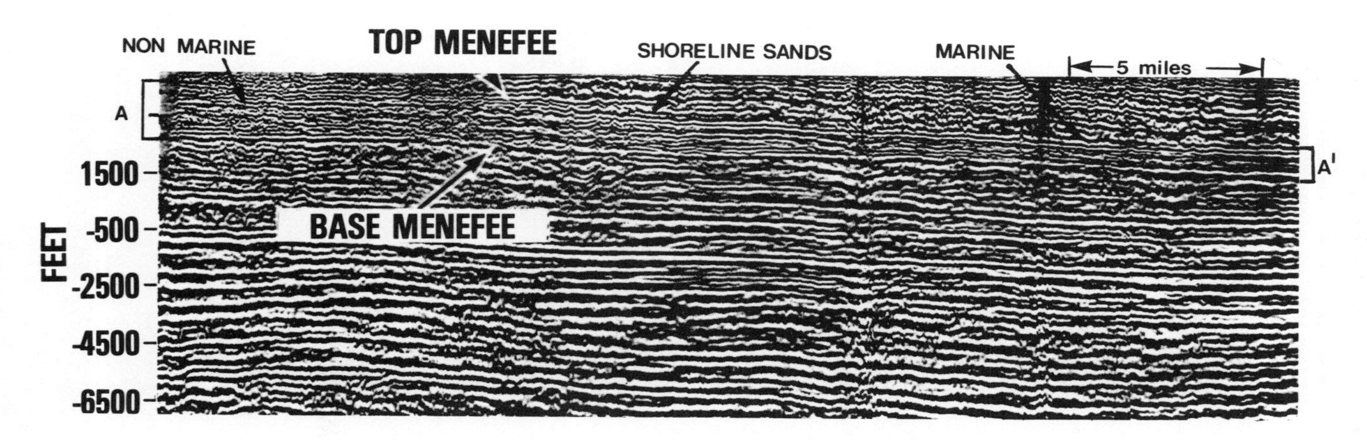

Figure 5.9. Seismic section from San Juan Basin. A number of wells along the line provide subsurface control. The top portion of the Menefee unit grades from nonmarine at the left through a shoreline region to marine at the right. AA′ marks the nonmarine sediments (the base of the unit at the right end is nonmarine). (From Vail, Todd, and Sangree, 1977; reprinted by permission of The American Association of Petroleum Geologists)

amplitude reflections in the marine zone. Reflections within the unit prograde across the unit following the attitude of constant time lines.

Figure 5.10 shows variations of reflection character associated with changes in depositional environment for a broad, low-relief mound in a shelf environment.

Different Concepts of Marine Onlap

The differences between the Vail et al. and the Brown and Fisher concepts, which were discussed in chapter 3, affect the evaluation of the reservoir character of deep-sea sediments. Offlapping prograding patterns on the slope, according to Brown and Fisher, generally indicate a sustained supply of sediment and imply a river source, but they are usually fine-grained (sand-starved) because they are so far from the sediment source (excepting oblique patterns involving wave-dominated energy, which produces good sorting). Carbonate-shelf systems generally prograde in sigmoidal patterns. Onlapping fans on slopes result from turbidity flows derived from the adjacent shelf sediments; they may be of reservoir quality if the shelf sediments contained coarse-grained material. However, widespread onlap probably indicates slow, fine-grained deposition.

Figure 5.6 illustrates one of the basic differences between the Vail et al. and Brown-Fisher viewpoints. There is little disagreement between their offlap diagrams showing the effects of a gradual rise in sea level. However, Vail et al. believe that marine onlap is evidence for a lowering of sealevel whereas Brown and Fisher attribute it to a cessation in the influx of terrigenous sediments and the dominance of erosion of the continental shelf. Brown and Fisher generally give more significance to marine erosion. Marine erosion effects were not evident in the situation indicated in the lower

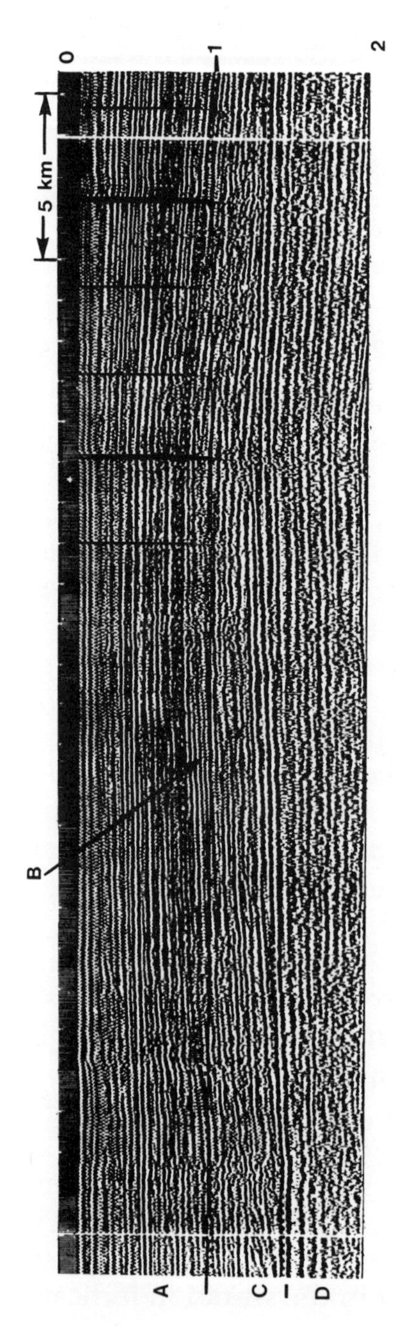

Figure 5.10. Broad, low-relief mound. A indicates delta-plain depositional environment, B delta-front, C prodelta, and D carbonate platform. (From Sangree and Widmier, 1977; reprinted by permission of The American Association of Petroleum Geologists)

left diagram of figure 5.6 because the influx of sediments and outbuilding of the shelf overwhelms erosion effects, which do not become evident until the influx of new sediment material ceases. Under the Brown-Fisher concept, the prospectiveness of marine offlap situations for stratigraphic traps is great, assuming that the shelf being eroded contained suitable reservoir rock material, because the marine onlap deposits tend to alternate with hemipelagic clays that seal reservoirs.

Classification Table for Clastic Facies

Sangree and Widmier provide a classification of the patterns for clastic facies, which is rephrased as table 5.1. This classification depends on the regional setting of units as well as their facies characteristics.

Brown calls seismic facies interpretation a "process of elimination" in which the interpreter eliminates lithofacies/depositional environments based on inconsistencies with what is known about the area, basin venue, relationships to other units, and so on, as well as reflection characteristics; thus one has only a few possibilities remaining when seismic facies analysis begins. In the interpretation process, special emphasis has to be given to lateral facies equivalents. An interpreter needs a good understanding of depositional processes, lithofacies composition, geometry, and spatial relationships.

The classification shown in table 5.1 has regional setting as the first-order division. Brown uses internal reflection configuration as the first-order division (table 5.2). The numbered points in table 5.2 provide entry into table 5.1 at the corresponding numbered points.

A number of items of seismic facies interpretation are not included in tables 5.1 and 5.2. These include:

Table 5.1. Seismic Facies Analysis

Regional Setting	Basis of Distinction	Subdivisions	Table 5.2	Interpretation	Other Characteristics
Shelf	Reflection character Unit shape: widespread sheet or gentle wedge Reflections generally parallel/divergent	High continuity, high amplitude	1	Platform deposits—alternating meritic shales/limestones	Possibly cut by submarine canyons
				Interbedded high/low energy deposits Shallow marine clastics mainly by wave transport Fluvial clastics/marsh interbedded	Distinguish on basis of location compared to other facies
		Variable continuity, low amplitude, occasional high amplitude	2	Fluvial/nearshore clastics, fluvial/wave-transport processes (delta platform)	Distinguish on basis of location compared to other facies
			4	Marine clastics, low energy turbidity currents or wave transport	Shale-prone if seaward of unit above Sand-prone if seaward of unit below
		Low continuity, variable amplitude	3	Nonmarine clastics, fluvial/marginal-marine transport	Occasional high amplitude and high continuity from coal members
	Mounded shape	Variable continuity and amplitude	10	Delta complex	Internal reflections gently sigmoid to divergent Occasional high amplitudes
		Local reflection void	8	Reef	See latter part of chapter 5

Table 5.1. Seismic Facies Analysis

Regional Setting	Basis of Distinction	Subdivisions		Table 5.2	Interpretation	Other Characteristics
Shelf margin—prograded slope	Internal reflection pattern (see figure 5.5)	Oblique fan-shaped multiple fans		5	Adequate sediment supply Shelf margin—deltaic High energy deposits in updip portions Occasionally due to strong currents in deep water	Moderate to high continuity and amplitude, reflections somewhat variable. Foreset (clinoform) dips to 10° (averaging 4–5°); steeper dips are calcareous Often fan-shaped to multiple fans
		Sigmoid elongate lens/fan		6	Low sediment supply Low depositional energy	High continuity, high to moderate amplitude, uniform reflections
Basin slope, basin floor	Overall unit shape (see figure 5.8)	Drape	Sheet drape	7	Deep marine hemipelagic; mainly clays, low energy	High continuity, low amplitude Drapes over preexisting topography
		Mounded	Contourite		Deep, low energy	Variable continuity and amplitude
			Fan-shaped		Variable energy, slump/turbidity currents	Discontinuous, variable amplitude At mouth of submarine canyons Composition depends on what was eroded up above
		Fill	Slope front fill	11	Low energy, deep marine clays and silts	Variable continuity and amplitude Fan-shaped to extensive along slope

Table 5.1. Seismic Facies Analysis

Regional Setting	Basis of Distinction	Subdivisions	Table 5.2	Interpretation	Other Characteristics
		Onlapping fill		Low-velocity turbidity currents	High continuity, variable amplitude
		Mounded onlap fill	13	High-energy turbidity currents	Discontinuous, variable amplitude
		Chaotic fill		Variable-energy turbidites	Overall mound in a topographic low, gouge common in base Discontinuous, variable amplitude
		Canyon fill	12	Variable superimposed strata Coarse turbidites to hemipelagic	Variable continuity and amplitude Statigraphy often self-trapping

Source: After Sangree and Widmier, 1977

Table 5.2. Seismic Facies Analysis

Internal Reflection Configuration	Possibilities	Table 5.1	Characteristics
Parallel/divergent	Shelf platform alternating neritic shales/limestones	1	High continuity, high amplitude Uniform subsidence Upper surface may be eroded
	Delta platform sandstone/shale	2	Low/moderate continuity and amplitude except high amplitude for coals
	Alluvial plain/fan delta (meander belt, channel fill)	3	Discontinuous, variable amplitude
	Basin floor; hemipelagic shale/silt/clay	4	High continuity, low amplitude Drapes over preexisting topography
Progradational	High sediment supply; clay/silt/sand	5	Oblique Variable continuity and amplitude On shelf or beyond shelf edge
	Low sediment supply; uniform conditions; calcareous clay/silt	6	Sigmoid High continuity, moderate uniform amplitude
Mounded/draped	Deep marine hemipelagic clay	7	Draped High continuity, low/moderate amplitude
	Biogenic carbonate reefs	8	Shelf/platform Elongate/subcircular shape Reflection free
	Canyon fan slump/turbidites	9	Slope/basin, fan-shaped Discontinuous, variable amplitude

Table 5.2. Seismic Facies Analysis

Internal Reflection Configuration	Possibilities	Table 5.1	Characteristics
Onlap/fill	Coastal onlap (paralic) delta/alluvial	10	Landward of shelf edge Low continuity, variable amplitude
	Marine onlap	11	Beyond shelf edge
	Canyon fill	12	In canyon
	Turbidite deposits	13	Basin floor

Source: After Brown and Fisher, 1979

(1) Hemipelagic deposits are generally concordant at the base.

(2) Divergent reflections indicate either differential subsidence, faulting, or diapiric flow contemporaneous with deposition.

(3) The formations underlying a shelf platform are generally sigmoidal whereas those underlying a delta platform are usually oblique.

(4) Delta platforms that extend into deep water usually involve growth faults.

(5) Prodelta sediments indicate water depths less than 100 m; in deeper water, deposits slump or are involved in density flow, or are otherwise reworked.

(6) Canyon-fill deposits are usually turbidite deposits interbedded with hemipelagic clays, so that they provide natural stratigraphic traps. The coarsest turbidities are usually near the base.

Reef Patterns

Reefs can be of a number of different types, as illustrated in the diagram of figure 5.11: patch reefs, shelf-margin reefs, pinnacle reefs and barrier reefs. Carbonate platforms and other carbonate sediments are also sometimes considered part of reefs; "reef" does not necessarily imply a bioherm built by coral or other organisms.

Shelf-margin and barrier-type reefs are distinguished by their elongate shape. They usually separate different depositional environments, often a quiet, shallow lagoon from deeper-water deposits. They can often be distinguished by different reflection patterns, so that the lateral change in reflection pattern aids in locating them. The reefs themselves

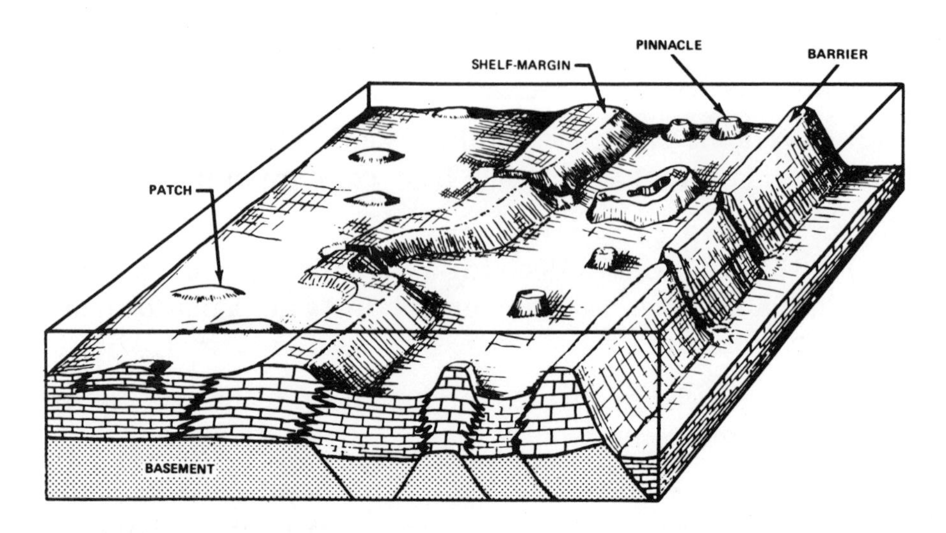

Figure 5.11. Types of carbonate build-ups. (From Bubb and Hatlelid, 1977; reprinted by permission of The American Association of Petroleum Geologists)

are often so irregular they do not produce coherent reflections, although frequently the top of the reef or sediments draped across the top produce a reflection that outlines the reef. Sometimes reef edges or the termination of reflectors against reefs produce distinguishable diffractions. The shelf edge or fault corner that provides the locale on which the reef grows can sometimes be seen. Shelf-margin and barrier-type reefs are shown in figures 1.8, 1.10, 2.9, 4.2, 5.12, and 5.13.

Patch and pinnacle-type reefs are generally smaller and more difficult to detect on seismic data. The primary key to finding them is knowing where to look; that is, knowing the portion of the section where conditions for their growth were apt to be propitious. Patch and pinnacle reefs are shown in figures 5.14 and 5.15.

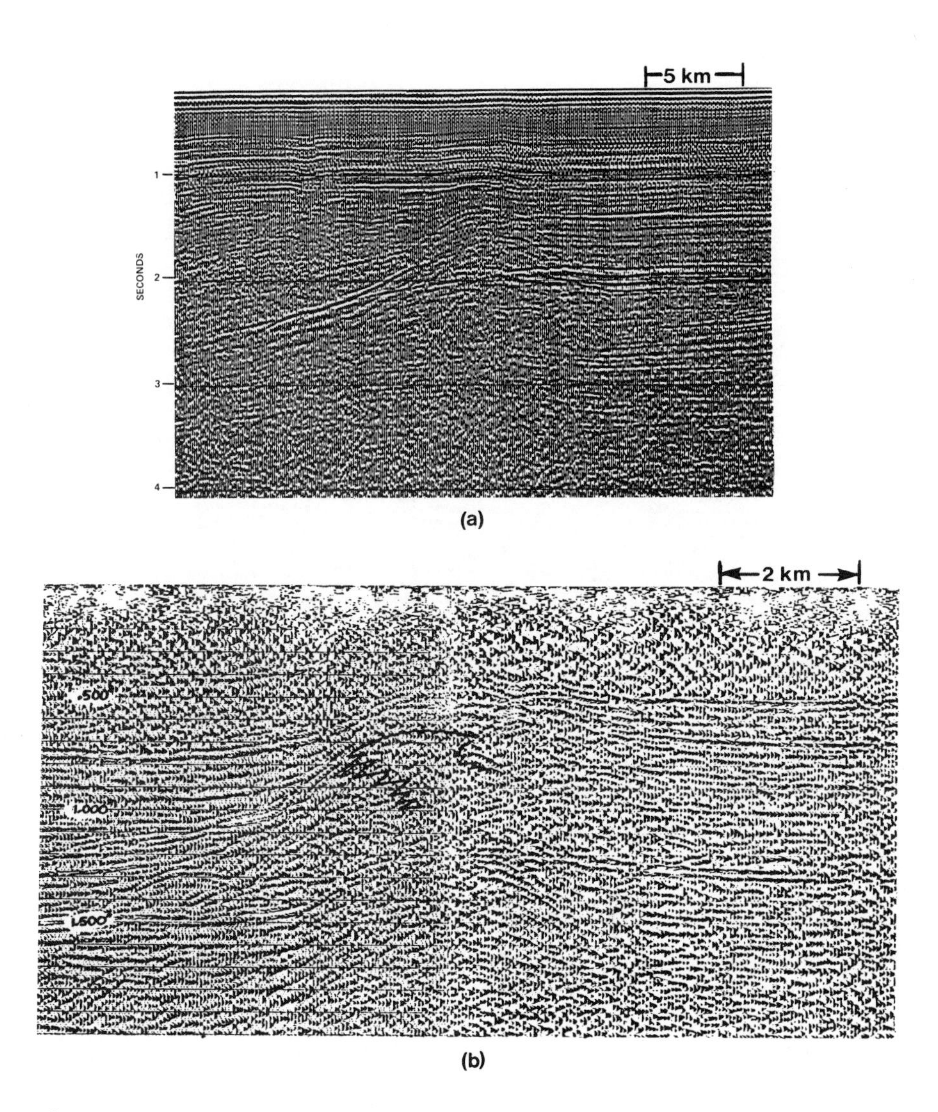

Figure 5.12. Seismic sections over reefs. (a) Line offshore West Africa. (b) Line across Golden Lane reef in Mexico; the reef may be larger than has been sketched in. (From Bubb and Hatlelid, 1977; reprinted by permission of The American Association of Petroleum Geologists)

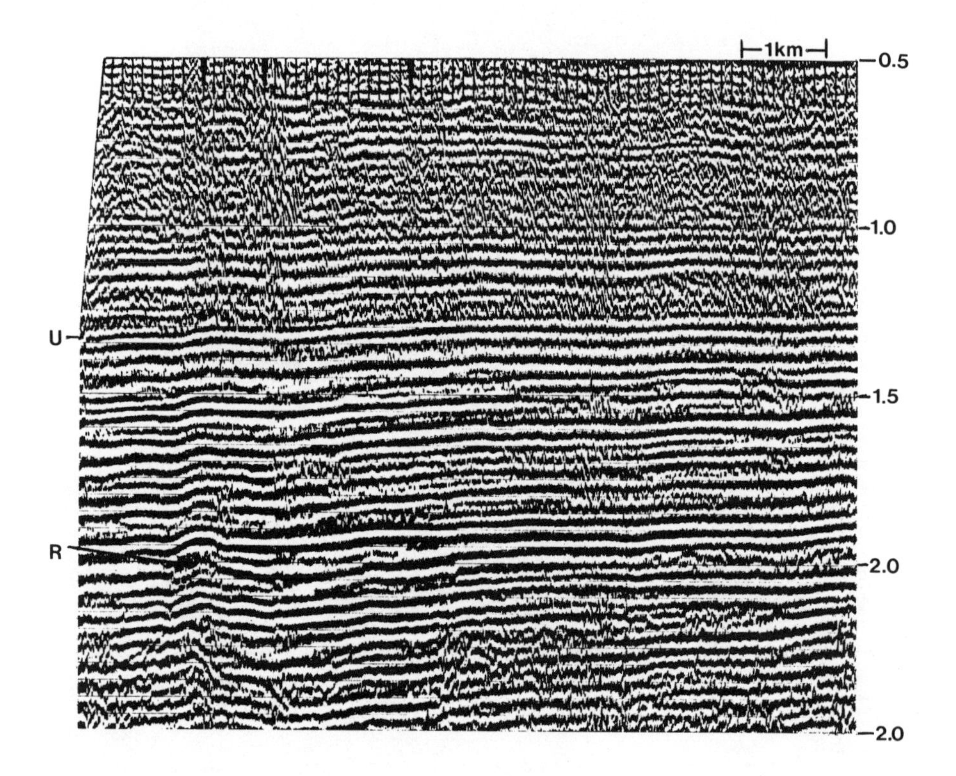

Figure 5.13. Phase section over a reef in Western Canada, R = reef. Note differential compaction over the reef and velocity uplift underneath. Differential compaction effects were sufficient to have significantly affected the sediments laid down on the unconformity U (there are also other unconformities in the section). (Reprinted by permission of Seiscom Delta)

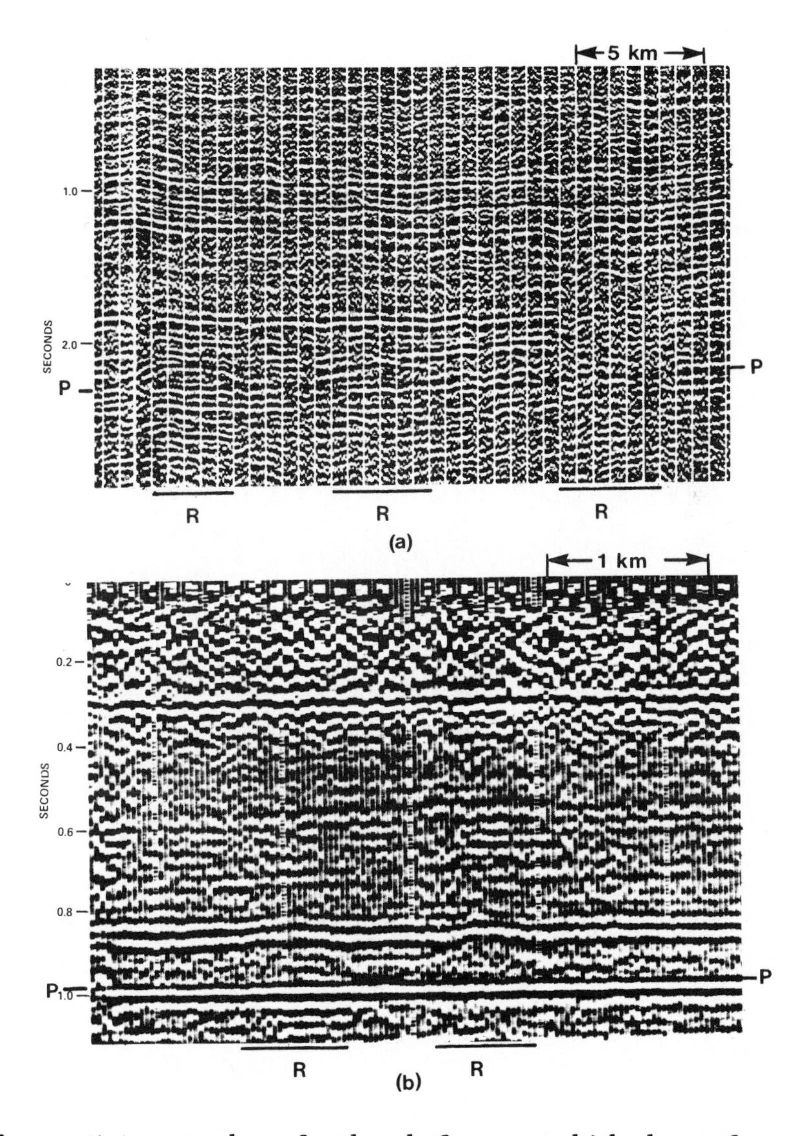

Figure 5.14. Patch reefs. The platform on which the reefs grew is labeled P and the reefs are above the labeled R. (a) Three African reefs, the one to the right having grown appreciably higher than the two to the left and (b) two small Canadian reefs. (From Bubb and Hatlelid, 1977; reprinted by permission of The American Association of Petroleum Geologists)

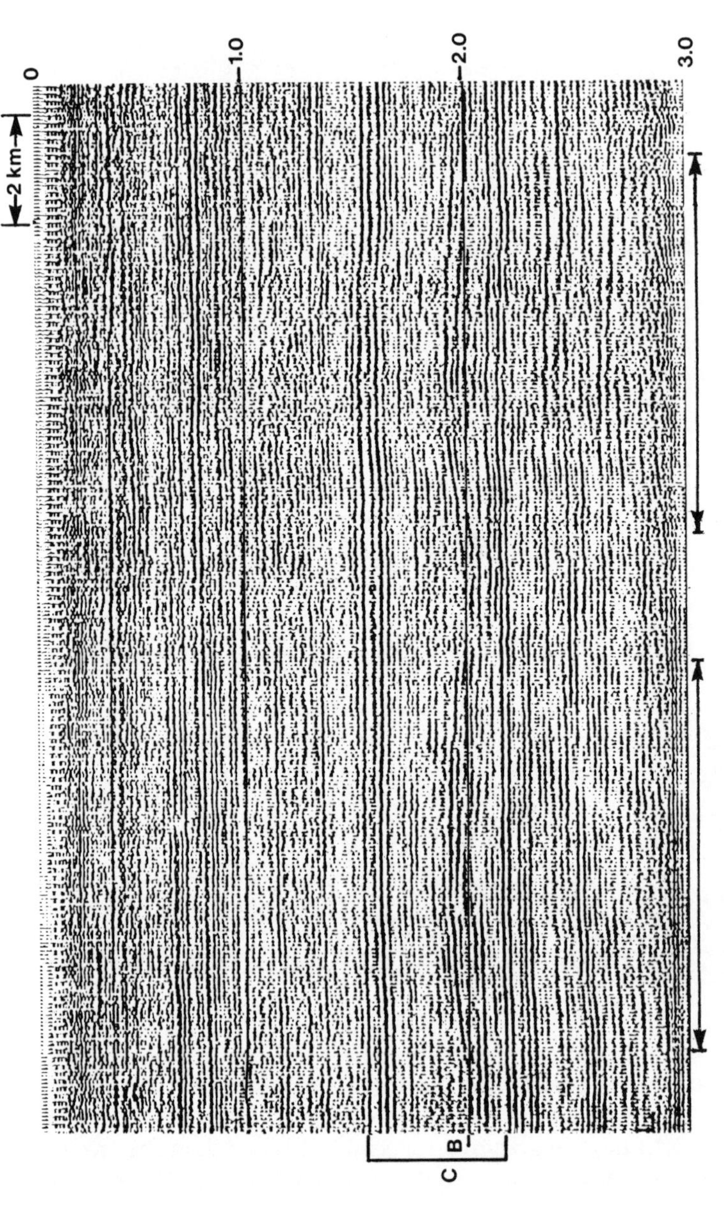

Figure 5.15. Two patch reefs in Etosha Basin of Southwest Africa. C denotes the carbonate portion of the section, B the base of the reefs; the region of the reefs is indicated approximately by the arrows below the section. The reef to the left has about 85 ms thickness (210 m), the one to the right 120 ms (300 m). (Reprinted by permission of Teledyne Exploration.)

Distinguishing Characteristics of Reefs

Figure 5.16 is a summary of criteria for distinguishing reefs on seismic sections. These criteria include:

(1) Reflections that partially outline reefs.

(2) Reflection voids, distinguishable by the sharp termination of reflections that onlap the reef.

(3) Changes in amplitude, frequency, or continuity of reflections at reef edges.

(4) Differences between reflection patterns on one side of the reef compared to those on the other; this is often especially marked for shelf-margin and barrier reefs where the back-reef and fore-reef patterns may differ markedly.

(5) The presence of diffractions and other types of events that mark reef edges.

(6) Differential compaction effects that produce a drape in the sediments over the reef; this is generally due to the off-reef sediments being more compactable than the reef itself, but occasionally the porous parts of a reef collapse and produce compaction effects. Compaction effects usually become gradually less with distance above the reef.

(7) Velocity anomaly for reflections underneath the reef (see below and figure 5.17).

(8) Location where reef growth should be propitious, such as on a hinge line, at the edge of a shelf, or on the uptilted edge of a fault block.

(9) Regional factors, such as the knowledge that the climate or environment associated with a particular reflection was propitious for reef growth.

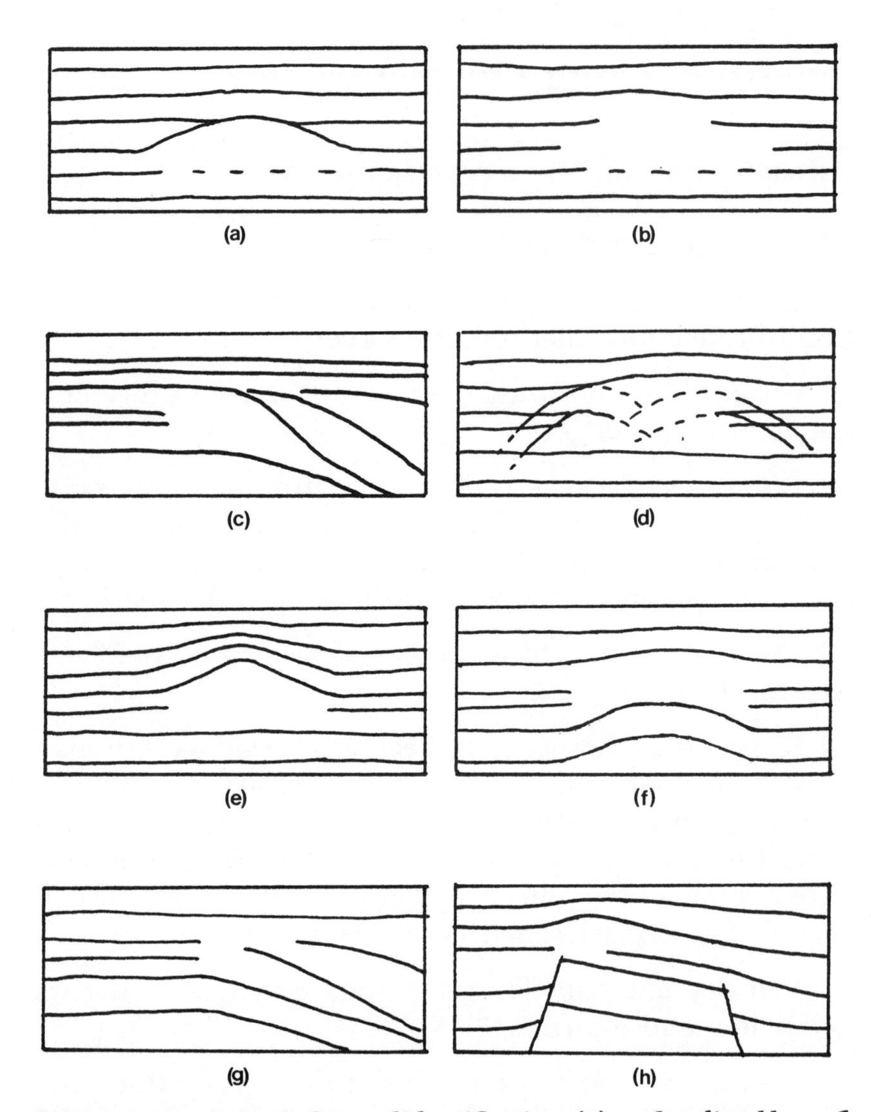

Figure 5.16. Criteria for reef identification: (a) reef outlined by reflections; (b) indicated by reflection void; (c) change in reflection pattern on opposite sides of reef; (d) diffractions from reef edges; (e) differential compaction over reef; (f) velocity anomaly underneath reef; (g) reef located on hingeline; and (h) reef located on structural uplift. (After Bubb and Hatlelid, 1977; reprinted by permission of The American Association of Petroleum Geologists)

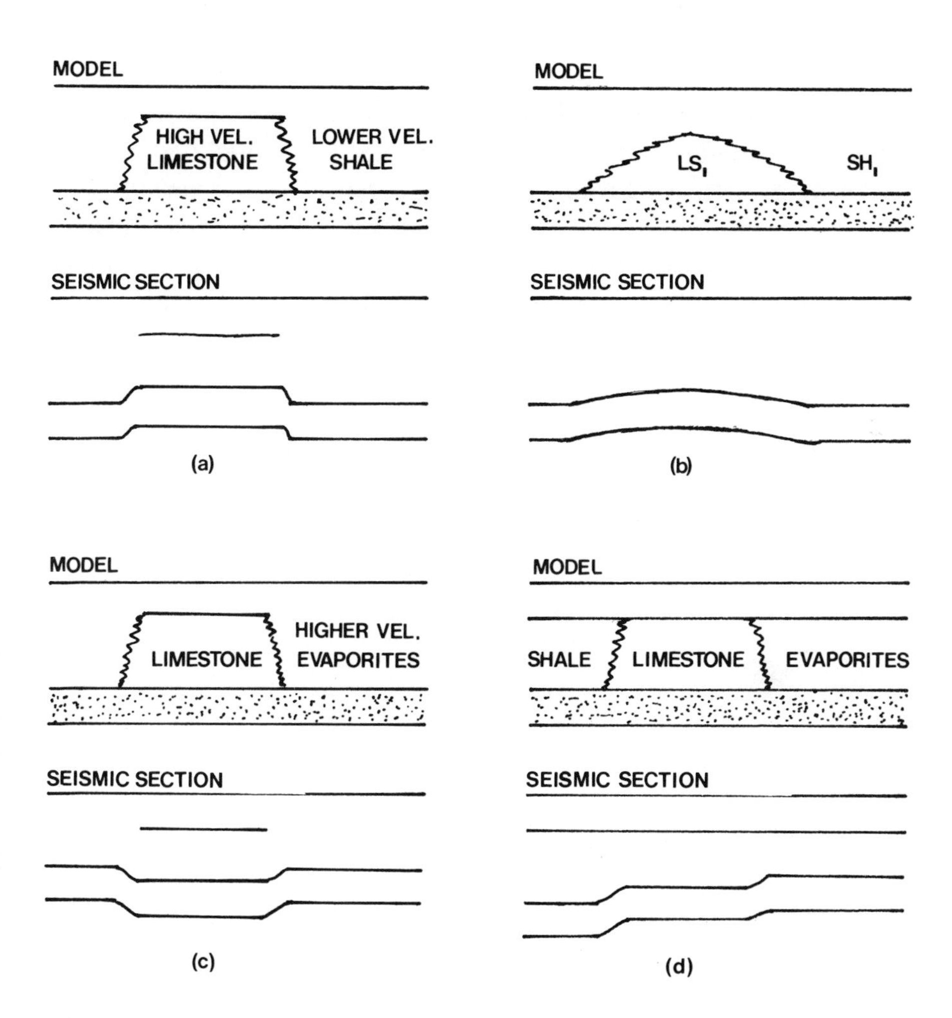

Figure 5.17. Velocity anomalies resulting from reefs: (a) reef surrounded by shale usually produces an apparent uplift in deeper reflections; (b) similar situation except reef is more mound-like; (c) reef surrounded by higher-velocity evaporites such as anhydrite produce an apparent depression in deeper reflections; and (d) reef might have lower-velocity sediments on one side, higher-velocity sediments on the other side, producing opposite types of velocity anomalies on opposite sides.

Reefs are usually evidenced by a combination of the foregoing, it being recognized that none are exclusive indicators of reefs.

Commonly, a reef has higher velocity than the surrounding sediments (see figures 5.17a and 5.17b), especially where the surrounding sediments are shales, and hence reflections under a reef usually show an apparent but unreal uplift. Sometimes a reef is surrounded by evaporite deposits such as anhydrite, which have higher velocity than the reef, and then the velocity anomaly under the reef is an apparent sag (see figure 5.17c). Occasionally the back-reef and fore-reef sediments will be of sufficiently different kinds that velocity anomalies will differ for opposite sides of a reef (see figure 5.17d).

6
Velocity and Resolution

Velocity as a Diagnostic

Velocity measurements are often used to determine the nature of sediments, so it is appropriate to examine the sensitivity of velocity to those factors we wish to determine and also to other factors that affect the reliability of determinations.

Figure 6.1 summarizes the effects of various factors on velocity. The decrease in velocity with increasing porosity is usually marked. While velocity increases with density, the range of density values is small. Velocity decreases slightly with increase in temperatures and seems to increase slightly with grain size. The effect of hydrocarbon saturation is apt to be marked, especially for low gas saturation. Velocity is essentially invariant with respect to frequency over a very broad frequency range. Velocity decreases with interstitial fluid pressure and increases with overburden pressure, but the overall effect depends on the difference between these pressures. If pore fluid and overburden pressures are

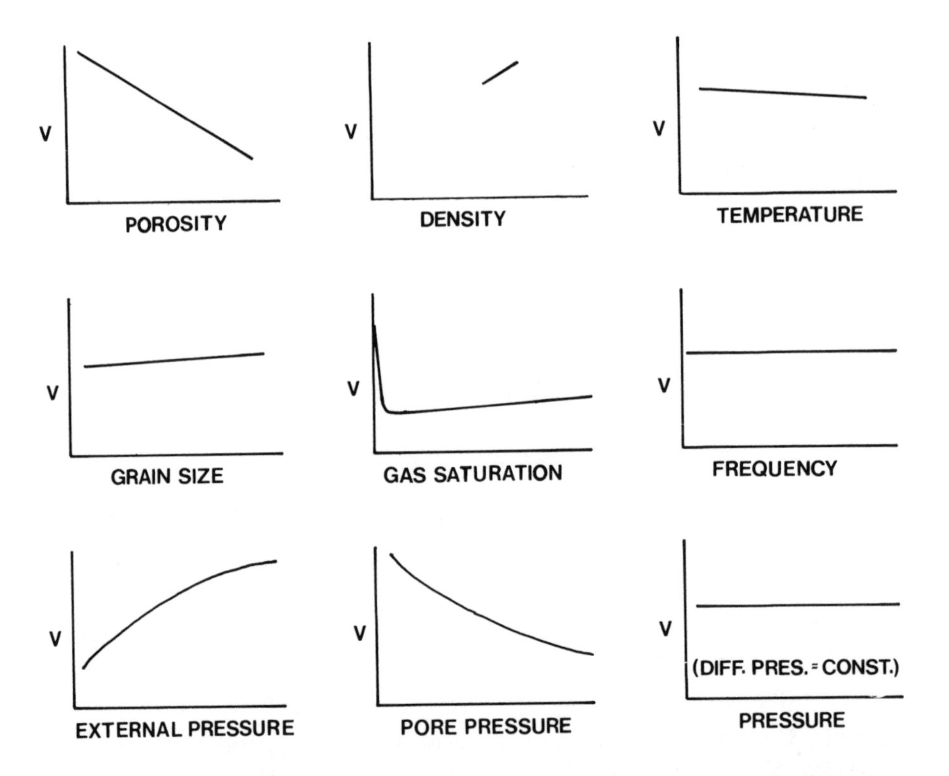

Figure 6.1. Effect of various factors on seismic velocity. (After Hilterman, 1977; reprinted by permission of The American Association of Petroleum Geologists)

changed by the same amount, the effects on the velocity cancel each other.

Of the various factors that determine seismic velocity, porosity is usually the most important. In clastic rocks, porosity depends on the differential pressure, the difference between the overburden and interstitial pressures. Porosity is reduced by increase of differential pressure in an irreversible process; thus the porosity of a clastic rock generally depends upon the maximum differential pressure during the rock's history.

When velocity is graphed for different rock types (figure 6.2) there is much overlap, that is, most rock types have an appreciable range of velocity. Except for very general rules, such as associating low velocities with clastic rocks and high velocities with carbonates or evaporites, velocity alone is generally not sufficient to distinguish rock type.

The principal reason for the broad range of velocity within any given rock type is the range of porosity (figure 6.3). High porosity values generally are associated with low velocity and low porosity values are generally associated with high velocity.

The porosity of a clastic rock (figure 6.4) generally decreases with compaction (depth of burial), but subsequent uplift does not change the porosity or the velocity significantly. Porosity generally becomes smaller as rock sorting becomes poorer and as cementation increases.

Figure 6.5 shows spontaneous potential (SP) and velocity traces from a well log. The SP is used to distinguish between sand and shale units. A best-fit curve drawn through the velocity of the shale units indicates lower velocity than a best-fit curve drawn through the velocity of the sand units. However, within each rock type, the variation is greater than the difference between rock types. Thus it may be possible to make a statistical analysis of sand-shale distribution while, for any individual sample, prediction based on velocity measurements is very uncertain.

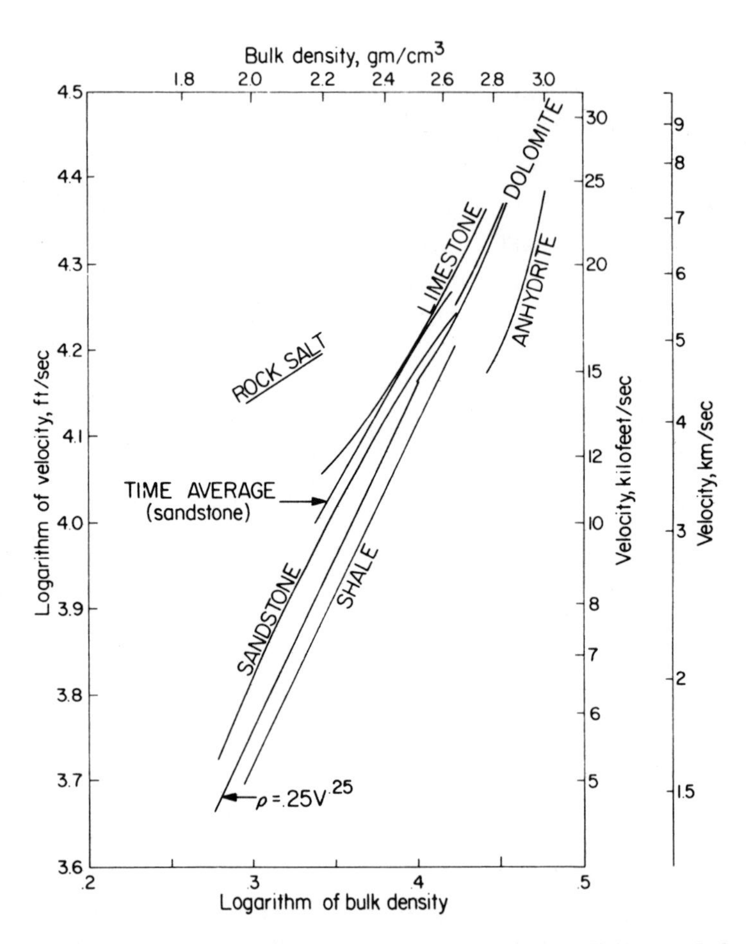

Figure 6.2. Velocity-density relation in rocks of different lithology. The graph also shows the time-average and Gardner relations. The time-average equation is $1/V = \phi/V_f + (1 - \phi)/V_m$ where V is a rock's velocity, ϕ its porosity, V_f the velocity of the fluid in the rock's pore space, and V_m the velocity of the rock's matrix material. This equation is an empirical one relating to a two-component system. Gardner's relation, $\rho = kV^{1/4}$ relates density ρ to velocity V using a constant k. This empirical equation is often used in synthetic seismogram and seismic log manufacture (chapter 7) when density information is not available. (From G.H.F. Gardner, L.W. Gardner, and Gregory, 1974; reprinted by permission of The Society of Exploration Geophysicists)

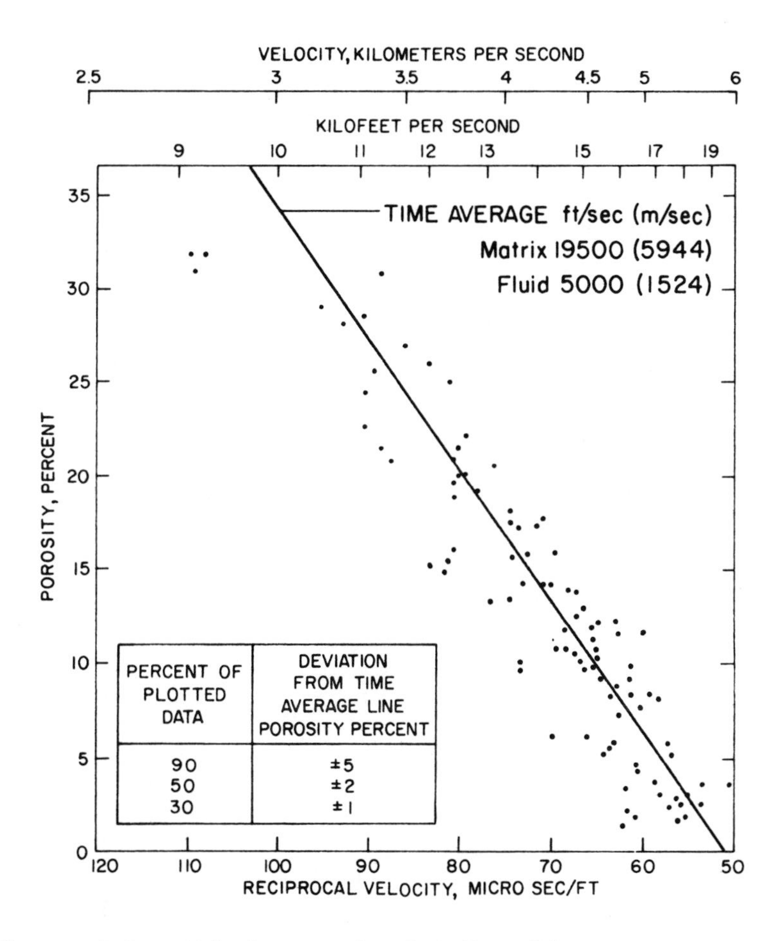

Figure 6.3.. Velocity-porosity data from laboratory measurements. The best-fit line is usually poorest at high porosities. (From Wyllie, Gardner, and Gregory, 1956; reprinted by permission of The Society of Exploration Geophysicists)

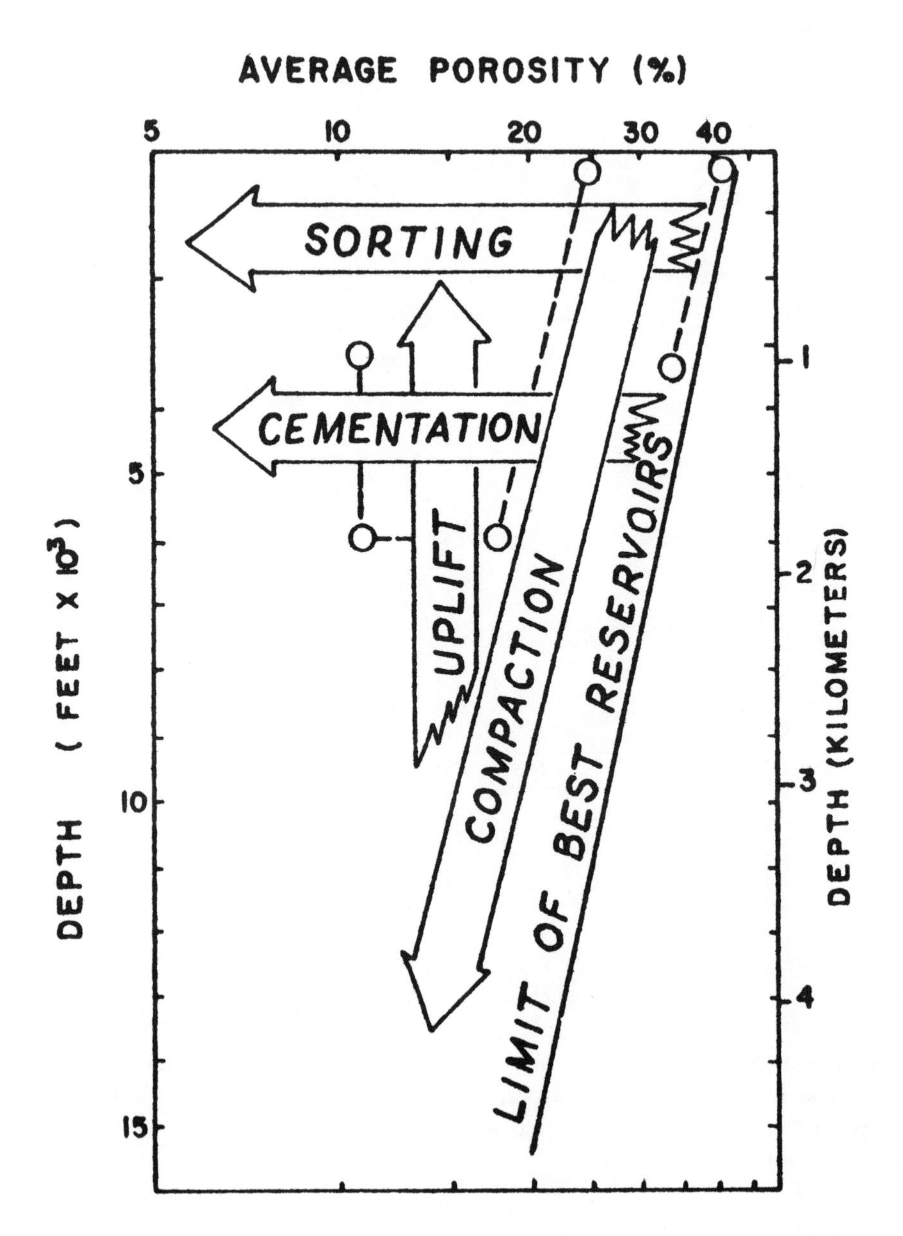

Figure 6.4. Effect of various factors on porosity. (From Zieglar and Spotts, 1978; reprinted by permission of The American Association of Petroleum Geologists)

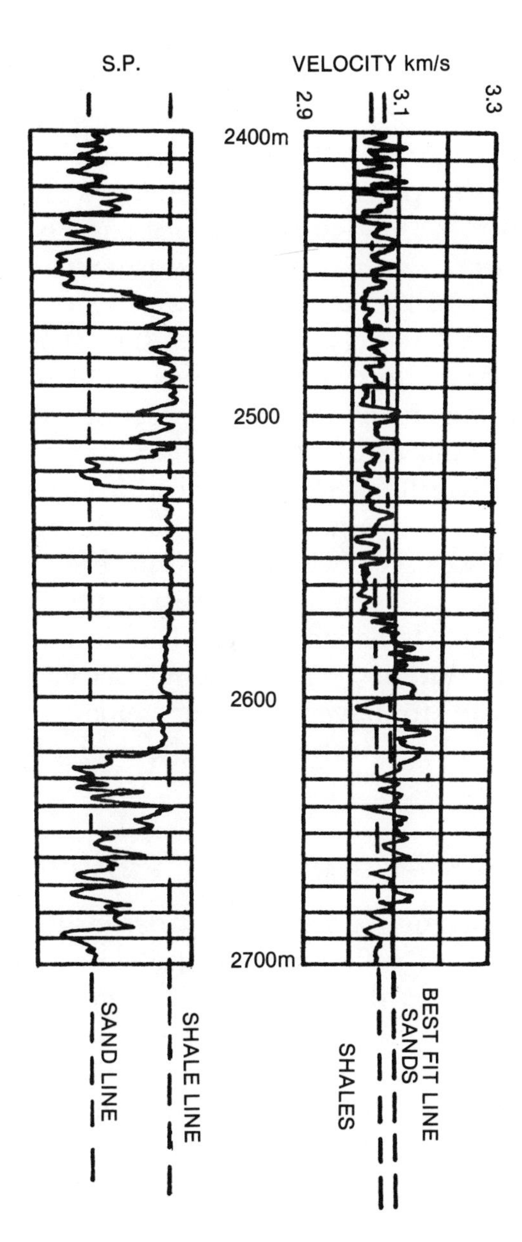

Figure 6.5. Velocity differences between sands and shales. (From Sheriff, 1978)

Figure 6.6, a portion of SP and velocity logs, shows a marked decrease in velocity associated with the top of a high-pressure shale section. The velocity decrease associated with abnormally high pressure is not always as sharp as indicated here because the pressure often changes more gradually. This association of velocity with pressure is often used to predict pressure. The nature of the interstitial fluid affects velocity. Figure 6.7 shows the lower velocity in the gas-bearing part of a sand. Gas has more effect than liquid (oil) in lowering velocity (figure 6.8a). Only a small amount of gas in the pore space produces a marked effect on velocity, but further increase in gas saturation has only a minor effect. Thus very small amounts of gas can produce large reflectivity (figure 6.8b) and, consequently, amplitude anomalies nearly as strong as those associated with commercial reservoirs.

Porosity is the most important stratigraphic feature that influences velocity. Shale content tends to lower velocity, lime and dolomite content to increase velocity. Thus a shaley limestone is apt to have a lower velocity than a sandy limestone, and a limey sand has higher velocity than a lime-free sand.

Measurement of Velocity

Velocity measurements are usually based on the change in the arrival time of reflections as the shot-to-geophone distance changes. This distance is called *offset*, the time difference because of offset is called *normal moveout*, the velocity that the normal moveout implies is called *stacking velocity* because it is what yields the optimum CDP stack, and the calculation procedure is called *velocity analysis*. Velocity analysis is a fairly expensive process, and hence an analysis is often run only every 1 to 3 kilometers and the intervening velocities are interpolated. Uncertainties in individual analyses sometimes produce fictious velocity anomalies (see also

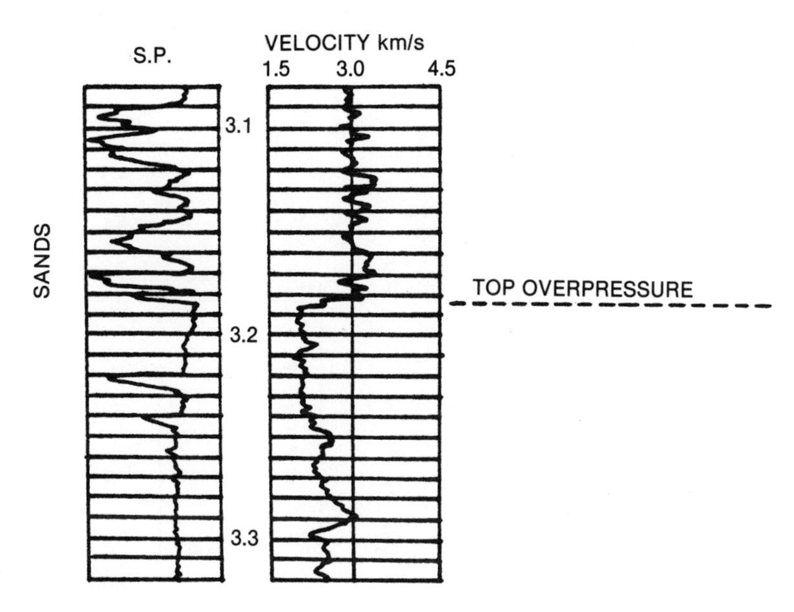

Figure 6.6. Velocity lowering because of excessive interstitial pressure. (From Sheriff, 1978)

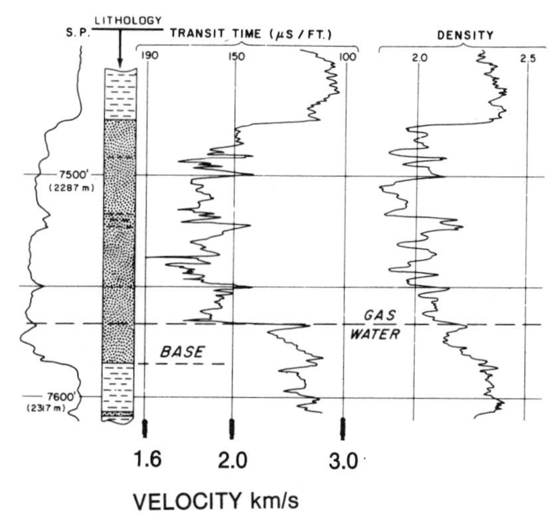

Figure 6.7. Sonic log through a gas-bearing sand, showing the lowering of velocity. (From Schramm, Dedman and Lindsey, 1977; reprinted by permission of The American Association of Petroleum Geologists)

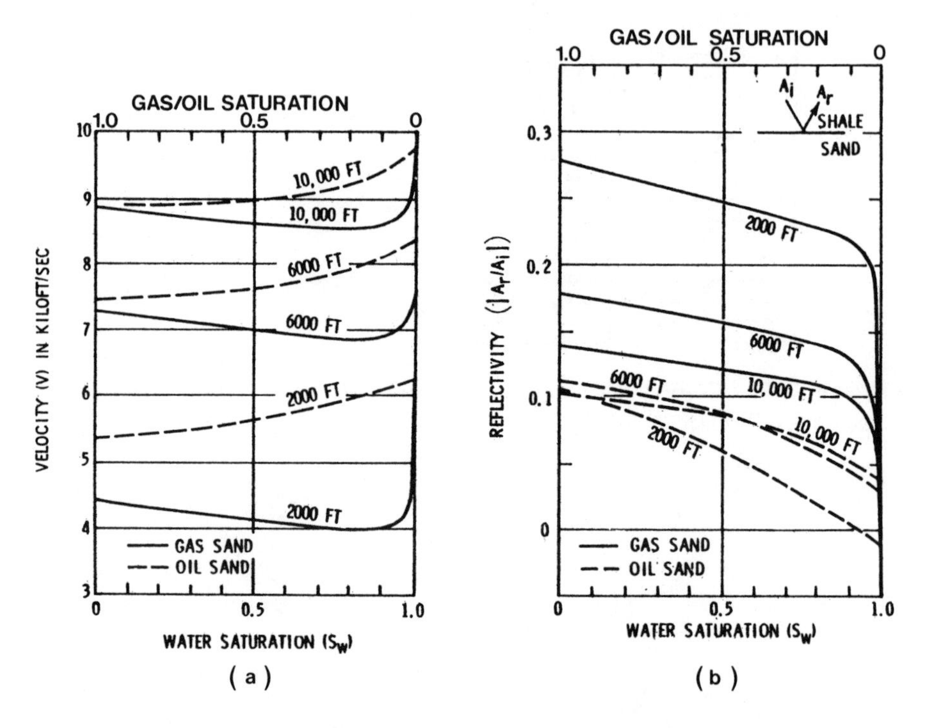

Figure 6.8. Effect of hydrocarbon/water saturation on velocity: (a) velocity versus saturation and (b) reflectivity versus saturation. (From Domenico, 1974; reprinted by permission of The Society of Exploration Geophysicists)

figure 2.6). To reduce the effects of noise, data over a small region are often averaged in making an analysis; this results in less scatter but then the measurements are averages that do not apply at specific locations. Velocity can also be calculated from amplitude information by trace inversion (see chapter 8).

Resolution

Resolution is defined as the minimum distance between two features so that one can tell that there are two features. Consider the reflections from a wedge that has a velocity intermediate between the velocities above and below the wedge (figure 6.9). When the wedge is thin, the reflection pattern is nearly the same as for a single interface. When the wedge thickness is greater than approximately 1/4 wavelength, the reflection pattern shows that at least two interfaces are involved. The minimum thickness of a bed in order to see the effects of the top and the base of the bed as distinctly separate is the *resolvable limit* and it is about 1/4 wavelength.

A minimum-phase wavelet was assumed in figure 6.9. A minimum-phase wavelet (figure 6.10a) has most of its energy concentrated towards the front of the wavelet, and most of the wavelets in actual exploration are nearly minimum phase. The wavelet shape most desirable for interpretation is zero phase (figure 6.10c); such a wavelet is symmetrical about some central point.

If the velocity above and below a wedge is the same (figure 6.11a), the reflection pattern is almost indistinguishable from that of a single interface unless the wavelet exceeds 1/4 wavelength in thickness. Figure 6.11b also assumes a minimum-phase wavelet, and figure 6.11c assumes a zero-phase wavelet; the resolvable limit is essentially the same. The changes in amplitude and their use in determining the thickness of

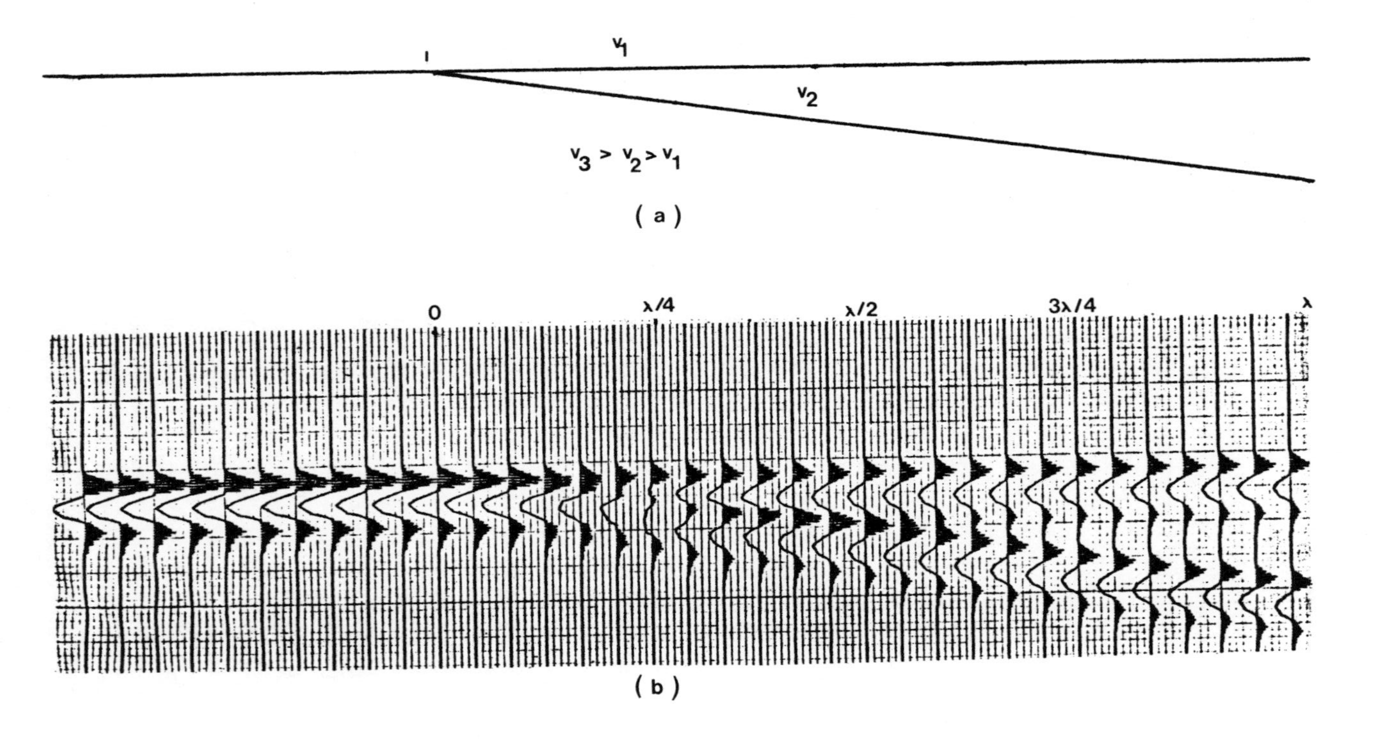

Figure 6.9. Reflection from a wedge. The thickness of the wedge is indicated as a fraction of the dominant wavelength. (a) Model. (b) Resulting seismic section. (From Sheriff, 1975; reprinted by permission of The European Association of Exploration Geophysicists)

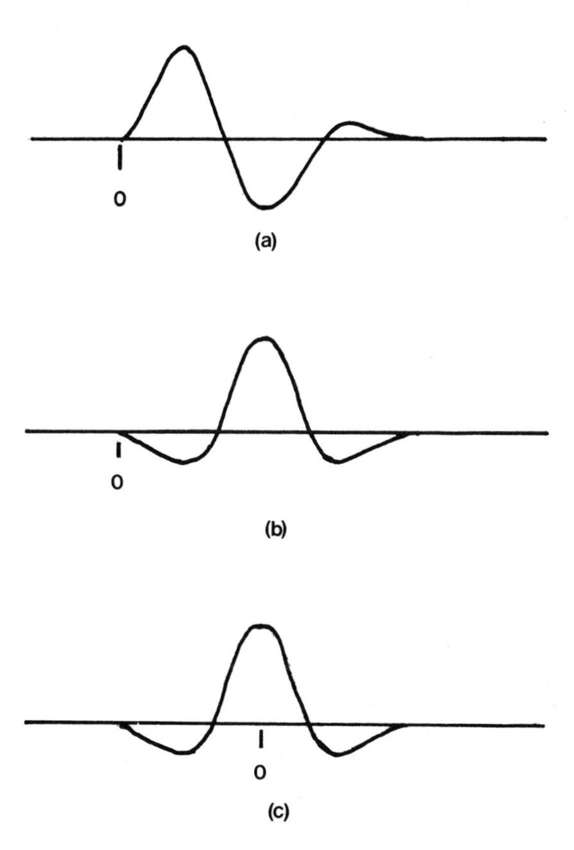

Figure 6.10. Diagram of minimum-phase and zero-phase wavelets. (a) Minimum-phase wavelet has energy build-up at front of wavelet. (b) A symmetrical causal wavelet of same length and frequency content. (c) Wavelet in (b) with time scale shifted; wavelet is now zero phase.

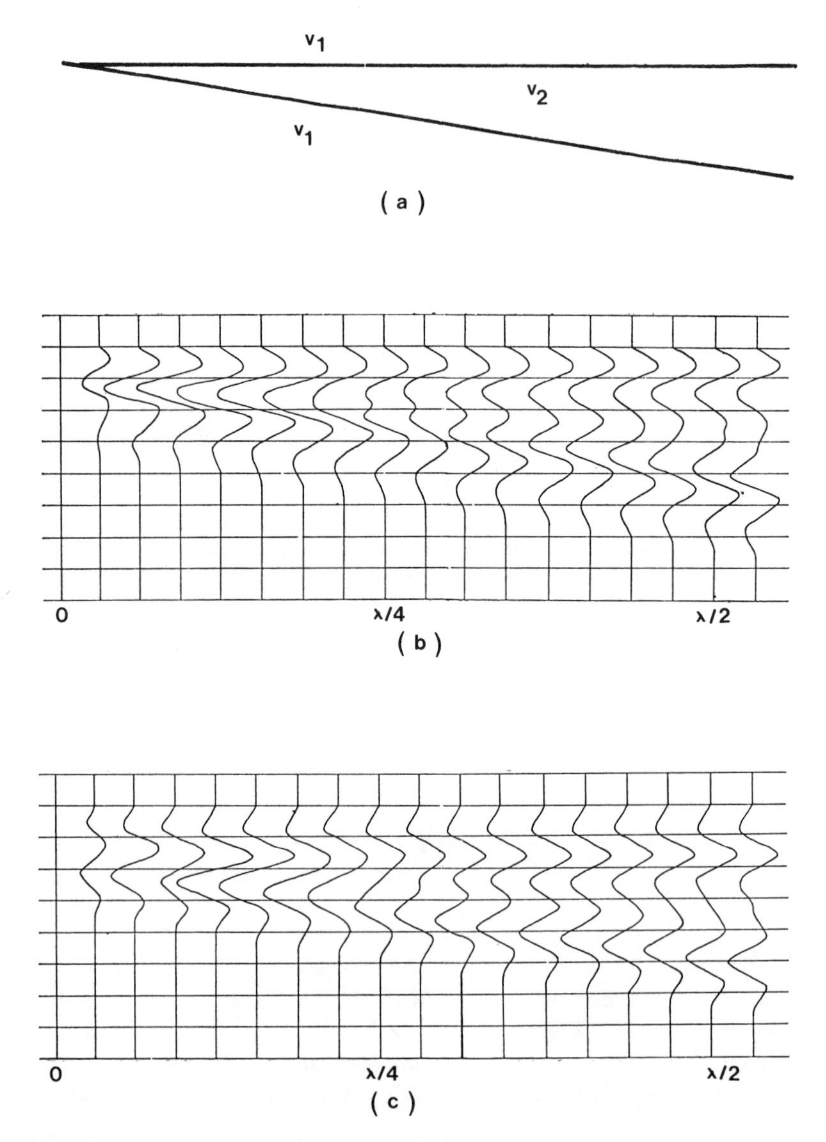

Figure 6.11. Reflection from a wedge immersed in a constant-velocity medium. Wedge thickness is indicated in wavelength units. (a) Model. (b) Response to a minimum-phase wavelet. (c) Response to a zero-phase wavelet. (After Sheriff, 1977a; reprinted by permission of The American Association of Petroleum Geologists)

thin beds is discussed further in chapter 8 (see also figure 8.6).

Note that the wedge in figure 6.11 generates a reflection even where its thickness is much smaller than the resolvable limit. The *detectable limit*, the minimum thickness for a layer to give a reflection, is of the order of 1/30 wavelength. Seismic frequencies are much lower than those used in well logging, as illustrated by figure 6.12, by a factor of about 100. Consequently, seismic resolving power is only one hundredth of that of well logging. Figure 6.13 is a nomogram relating velocity, frequency, and wavelength; it can be used to determine the resolvable- and detectable-limit criteria. In the shallow part of the earth, the velocity is generally small and the frequency is apt to be high, resulting in a wavelength of the order of 40 m, or resolvable limit of 10 m and detectable limit of 1.3 m. Deeper in the section velocities are higher (perhaps 5 km/s) and frequencies lower (perhaps 20 Hz), so wavelengths are of the order of 250 m, with resolvable and detectable limits of 62.5 m and 8.3 m respectively. Thus, there is nearly a 10:1 ratio between wavelengths shallow and deep in the earth.

Fresnel-Zone Effects

A factor in horizontal resolving power is the Fresnel zone, the portion of the reflecting surface from which energy returns to a geophone within a half-cycle after the reflection onset (figure 6.14). Energy reflected from this zone interferes constructively and builds up the reflection.

Figure 6.15 shows the reflections from reflectors of different lengths measured in terms of the Fresnel-zone size. Where the reflector is larger than about one Fresnel zone, the reflection shows the shape of the reflector whereas for small reflectors, the arrival time patterns are almost identical

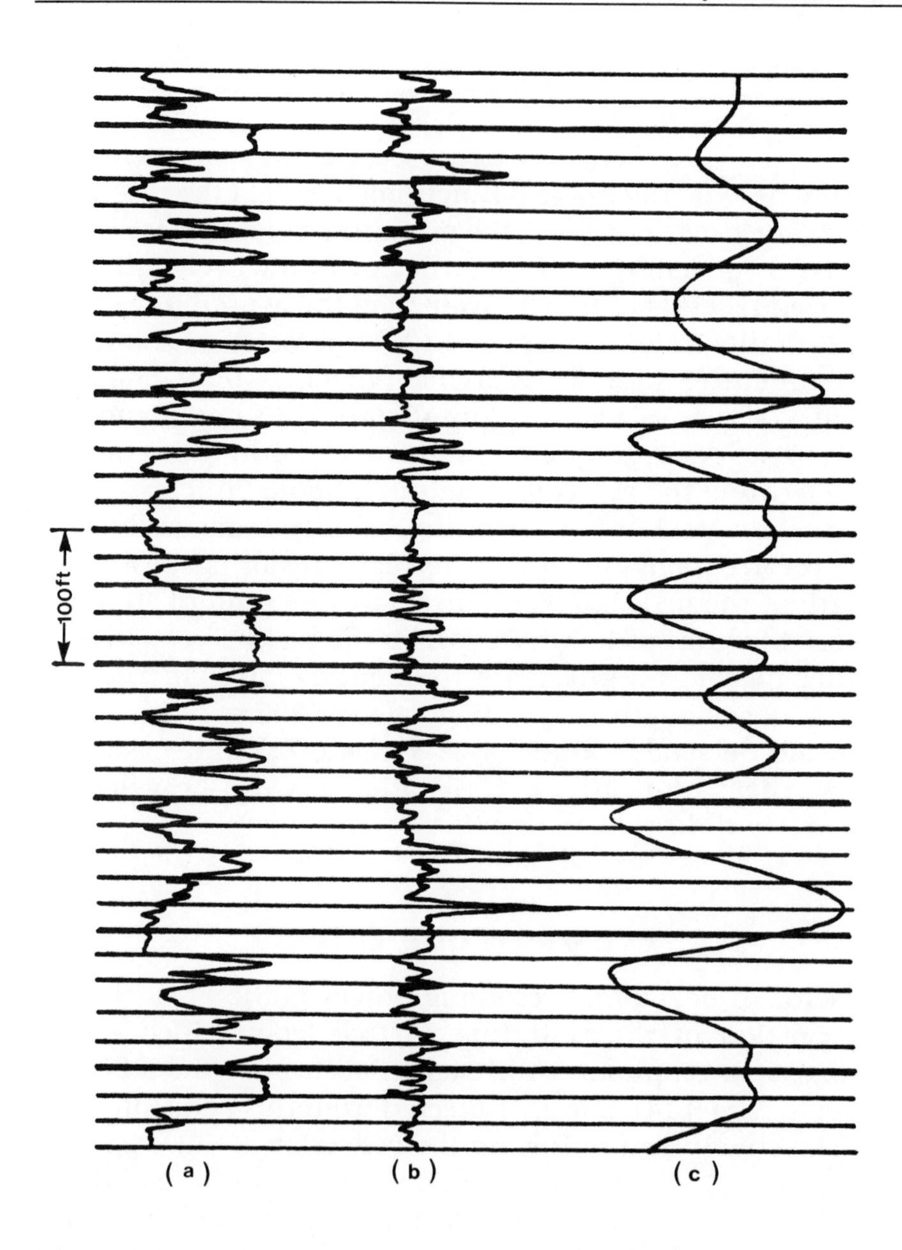

Figure 6.12. Frequency content of logs and seismic data: (a) spontaneous potential (SP) log; (b) velocity log; and (c) seismic trace. (Reprinted by permission of Chevron Oil Company)

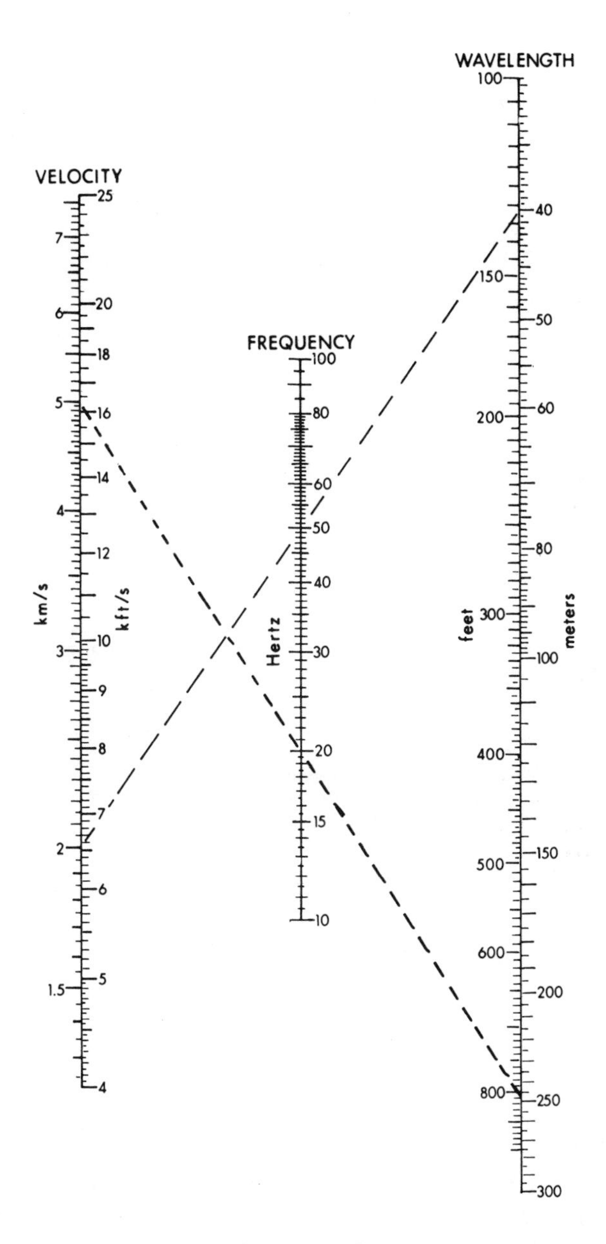

Figure 6.13. Nomogram relating velocity, frequency, and wavelength. Any straight line yields a correct solution. The two dashed lines show the examples given in the text. (From Sheriff, 1980; reprinted by permission of The Society of Exploration Geophysicists)

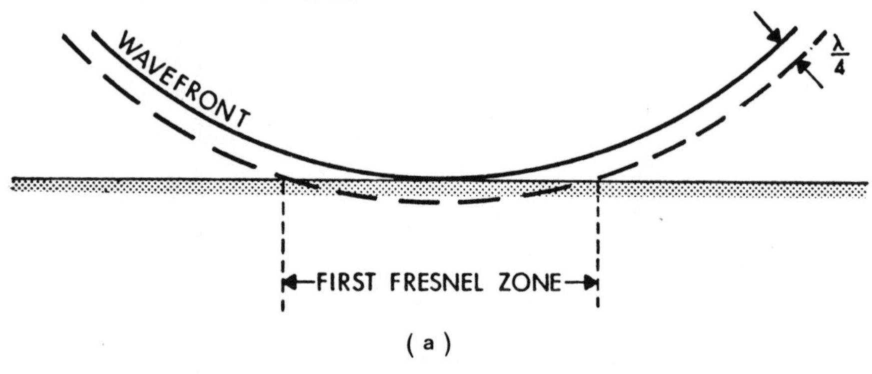

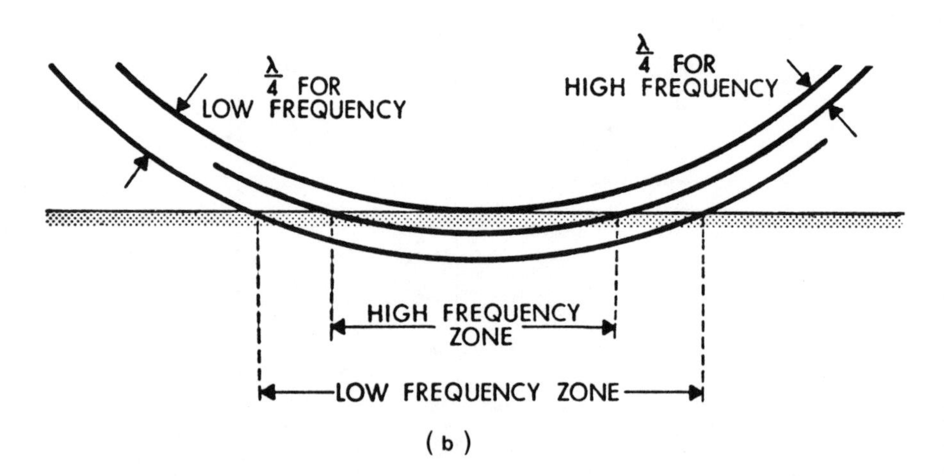

Figure 6.14. Fresnel zone. (a) The first energy to reach a geophone from a plane reflector is from the point where a wavefront is first tangent to the reflector; the area of the reflector from which energy can reach the geophone within the next half-cycle is limited by the circle that the wavefront a ¼-wavelength later makes with the reflector. (b) The Fresnel zone is larger for low-frequency components than for high-frequency ones. (From Sheriff, 1977a; reprinted by permission of The American Association of Petroleum Geologists)

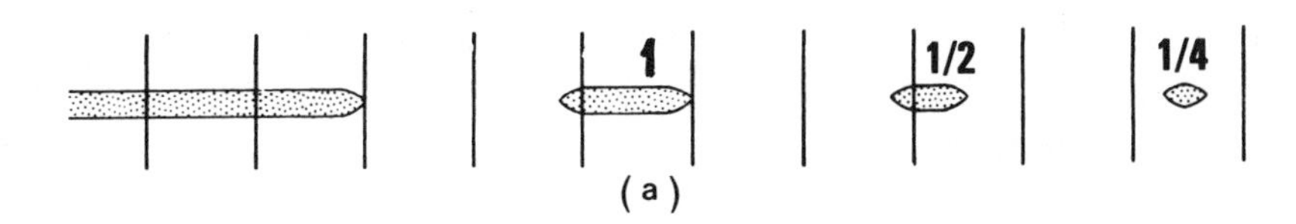

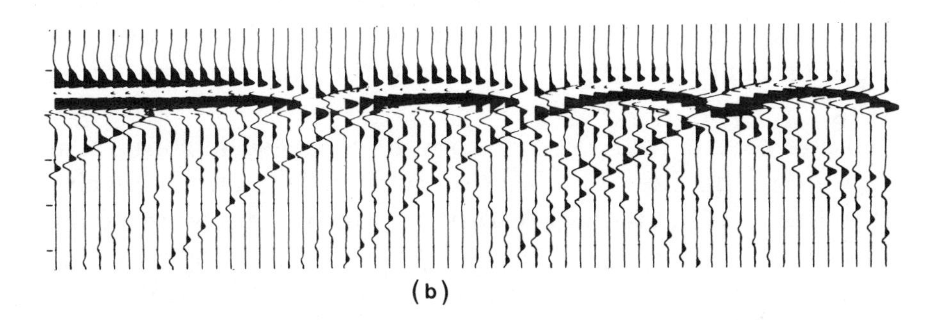

Figure 6.15. Reflections from reflectors of limited dimensions. (a) Cross section of model; vertical lines are spaced by the Fresnel-zone size (specific dimensions are not included because the effect is dimensionless). (b) Seismic section resulting from traverse over the model. The peak amplitude of the four reflections are, respectively, 100%, 87%, 55%, and 40%. (After Meckel and Nath, 1977; reprinted by permission of The American Association of Petroleum Geologists)

(but there are amplitude and other differences). The magnitude of the Fresnel zone is illustrated by the nomogram of figure 6.16. If one has a hole in a reflector, as shown in figure 6.17, the Fresnel zone laps onto the reflector surrounding the hole so that the reflection appears to be continuous through the hole. Such a hole might represent a pinnacle reef surrounded by horizontal bedding. There are, however, changes (such as in the amplitude) that can be used to distinguish the situation.

Another illustration of the effect of the Fresnel zone is shown by the box model of figure 6.18. A seismic line to the side of the box shows a reflection from the box (as well as the continuous reflection from the main reflector), because part of the Fresnel-zone region laps onto the box (see also figure 8.15).

To the extent that Fresnel-zone effects result from variations along the line of the section without changes occurring off to the side, horizontal resolution can be improved by migration (see chapter 2). However, while migration removes Fresnel-zone smearing, it substitutes migration noise (the migrating of nonreflection energy as if it were part of reflections) and it does not correct for out-of-the-plane effects.

Spatial sampling (the distance between geophone groups in seismic data acquisition and the areal extent of geophone groups) also limits horizontal resolution. The central points of reflector sampling are spaced at half the geophone group spacing and differences between these successive samplings have to indicate the reflector attitude. Where spatial sampling is too coarse, ambiguities in dip result; this effect is called *spatial aliasing.*

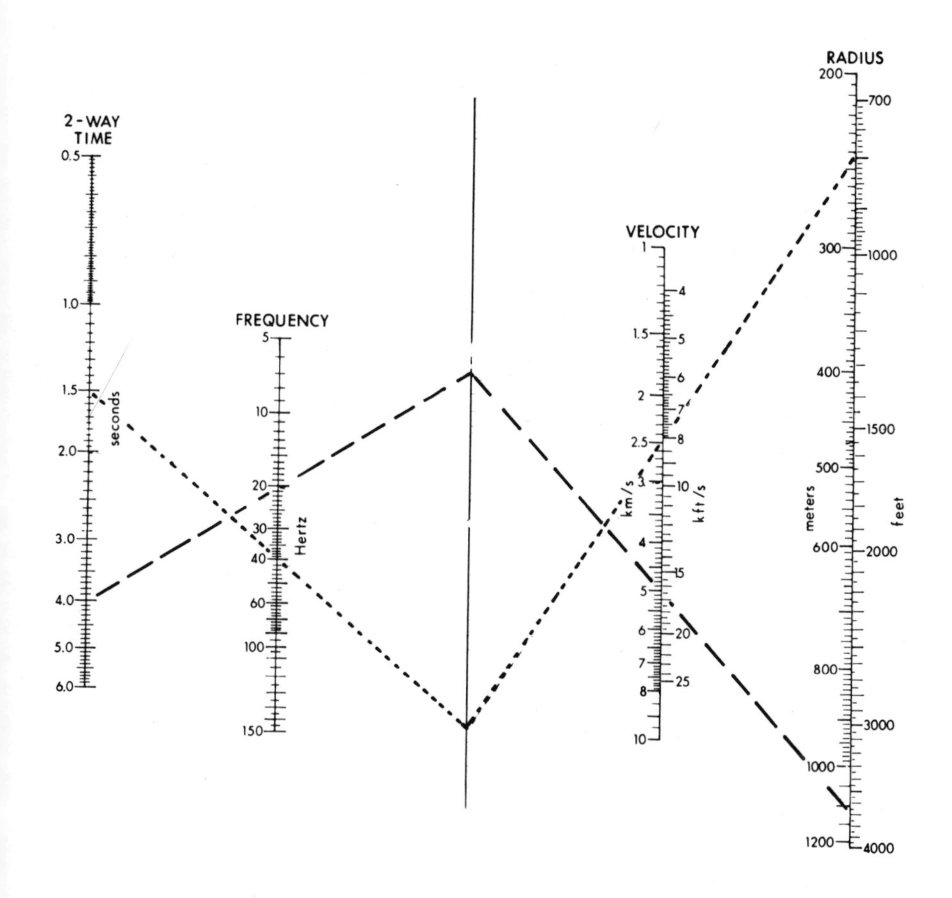

Figure 6.16. Nomogram for determining size of the Fresnel zone. A line connecting the arrival time and the frequency intersects the central line at the same point as a line connecting the average velocity and the radius of the Fresnel zone. For example, a reflection at 1.5 seconds with a 40 Hz component corresponds to a Fresnel-zone radius of 240 meters if the average velocity to the reflector is 2.5 km/s, and a reflection at 4.0 seconds with a 20 Hz component corresponds to a Fresnel-zone radius of 1,120 meters if the average velocity is 5 km/s. (After Sheriff, 1980; reprinted by permission of The Society of Exploration Geophysicists)

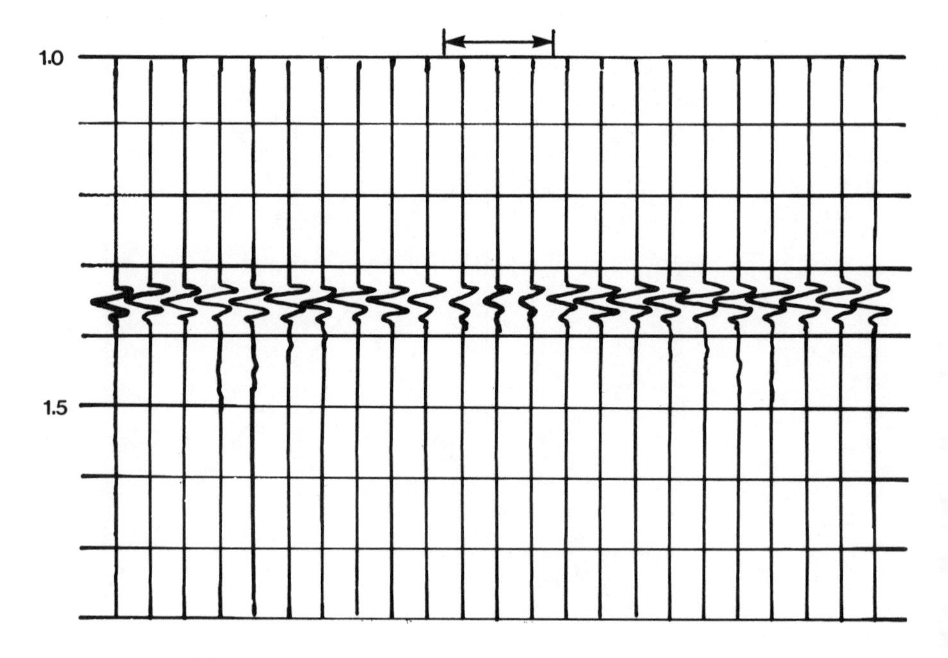

Figure 6.17. Reflection from a reflector containing a hole whose width is three trace intervals (indicated by arrow at top). (From Sheriff, 1977a; reprinted by permission of The American Association of Petroleum Geologists)

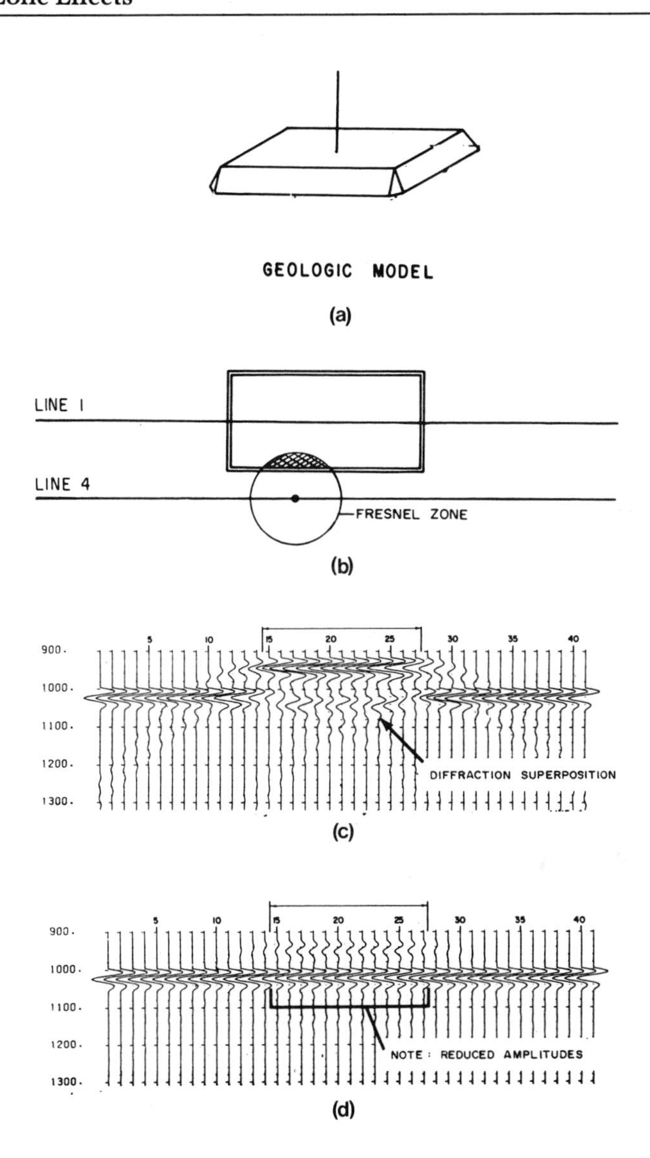

Figure 6.18. Reflections from a box model. (a) Isometric diagram of the model; the length: width: height: depth ratios are 10:5:1:10. (b) Plan view showing Fresnel-zone dimensions relative to box dimensions. (c) Line over the top of the box. (d) Line to the side of the box. (After Neidell and Poggiagliolmi, 1977; reprinted by permission of The American Association of Petroleum Geologists)

7
The Composition of a Seismic Trace: Modeling

The Modeling Concept

Modeling is the principal tool used in reflection character analysis. Both forward and inverse modeling are used (figure 7.1). *Forward or direct modeling* involves computing the effects of a model. It starts with a model of the lithology and calculates the seismic trace. It is used to see how stratigraphic changes affect seismic data and to test whether postulated stratigraphic changes provide feasible explanations of observed effects. *Inverse modeling* involves calculating a possible model from observation of effects. It starts with observed seismic traces and calculates acoustic impedance variations. It is used as an aid in interpreting trace-to-trace variations in terms of lithologic variations.

Modeling invariably involves a concept, an idea of a relationship between cause and effects. Sometimes it involves an actual physical model and the concept that effects will be scaled according to the ratio of model dimensions to actual

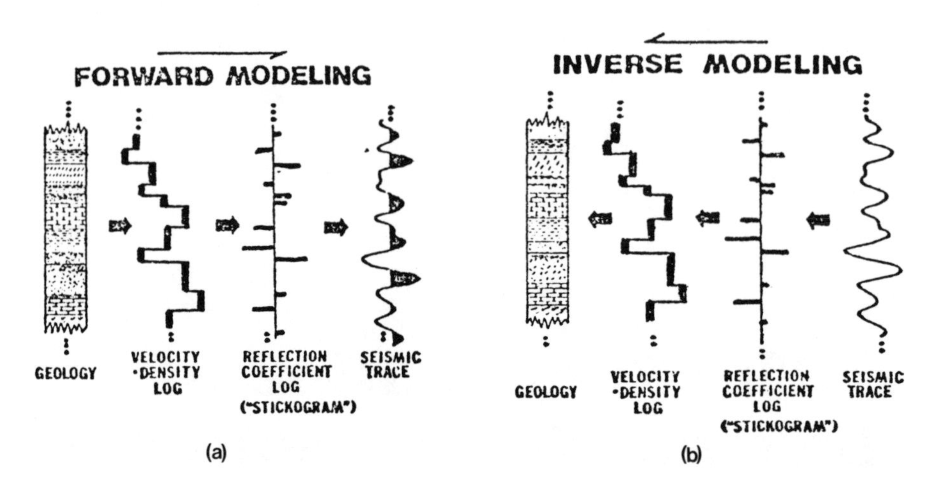

Figure 7.1. Forward and inverse modeling: (a) forward modeling starts with the geologic model and derives the seismic trace and (b) inverse modeling proceeds from the seismic trace toward the geology. (From Indonesian Petroleum Association Proceedings, 1978; reprinted by permission of Stommel and Graul)

dimensions. More commonly, modeling is done by manipulation of numbers in a digital computer according to a set of rules that incorporate the modeling concepts. Sometimes a model is only a mental concept that we use in thought processes.

Whatever the type of model, it will differ from the actual earth and from the processes we are modeling in many ways. The essence of a good model is that it includes proper consideration of the most important aspects without involving the complications of lesser-important aspects. However, judgment is required to separate important from unimportant considerations. What is important may depend on specific objectives, so that modeling may change as objectives change. Furthermore, there is no assurance of uniqueness in modeling since it is always possible that a different model or different modeling rules may give an equivalent result. Thus, achieving a desired result is not an adequate test of the validity of the model nor of the modeling program. On the other hand, while modeling cannot "prove" a point because of non-uniqueness, it can disprove, that is, it can show that a certain model cannot represent the situation that gives observed results.

Modeling involves incisive analysis as to what is or is not important, and thus provides one of the best ways to develop an understanding of which conclusions are reasonable and which are beyond our present abilities to distinguish. Its educational value for stratigraphic interpretation is tremendous.

The Convolutional Model

In stratigraphic modeling we are usually concerned with a distribution of rocks with different physical properties, and our objective is to see how this distribution affects the passage of seismic waves. We almost always use the *convolutional model* in our thinking. The convolutional model states that a seismic trace is the result of convolving the reflectivity function of the earth with a wavelet, with noise added. The reflectivity function is the waveshape we would record from the actual earth if the wavelet were impulsive, and the wavelet is the waveshape from a single reflector using the actual wavelet.

The physical picture that the convolutional model implies is of a downgoing wavetrain, the "wavelet," which is successively reflected from places at which the physical properties change. Figure 7.2 shows a sequence of changes in lithology, an associated graph of the physical property "acoustic impedance" (the product of density and velocity), and wavelets reflected at each change in acoustic impedance. The magnitude of a reflected wavelet is proportional to the magnitude of the change in acoustic impedance; the polarity of the reflected wavelet depends on the sign of the change. The superposition of all the individual reflections (plus noise) gives the seismic trace.

Reflectivity

When the angle of incidence at an interface is nearly zero, that is, when the approaching ray is perpendicular to the interface (normal incidence), the ratio of the amplitude of the reflected wave compared to the amplitude of the incident wave, called the *reflection coefficient* or *reflectivity*, is given by the following equation:

$$\begin{aligned}\text{reflection coefficient} &= \text{reflectivity}\\ &= \frac{\text{amplitude of reflected wave}}{\text{amplitude of incident wave}}\\ &= \frac{\text{acoustic impedance below} - \text{acoustic impedance above}}{\text{acoustic impedance below} + \text{acoustic impedance above}}\end{aligned}$$

$$= \frac{\text{change in acoustic impedance}}{\text{twice the average acoustic impedance}}$$

$$= \frac{\text{change in logarithm of acoustic impedance}}{2}$$

These equations can be expressed in symbols where ϱ = density, V = velocity, and the subscripts 1 and 2 indicate the values above and below the interface:

$$R = \frac{\varrho_2 V_2 - \varrho_1 V_1}{\varrho_2 V_2 + \varrho_1 V_1} = \frac{\Delta(\varrho V)}{2(\overline{\varrho V})} = \frac{\Delta \log \varrho V}{2}$$

where, the Δ signifies "change in" and the superscribed bar indicates average. R is the reflectivity at normal incidence. Most seismic reflection work involves angles of incidence that are sufficiently close to zero (within about ±20°) that this relationship gives a close approximation. The earth's reflectivity is simply the sum of the reflectivities of all the interfaces in the earth, separated by the two-way travel time between them.

Forward Modeling: Synthetic Seismogram Manufacture

Forward stratigraphic modeling is also called synthetic seismogram manufacture. We usually take the earth's reflectivity as derived from logs (or concepts as to what is probable) and convolve it with an equivalent wavelet. The wavelet shape used is our concept of what the downgoing wavetrain looks like; it is often derived from actual data. The synthetic seismogram that results is then compared with an actual seismic record (figure 7.3). If the match is poor, we conclude that either the earth's reflectivity or the equivalent wavelet shape needs to be changed. We usually make the changes we feel are appropriate and repeat the procedure,

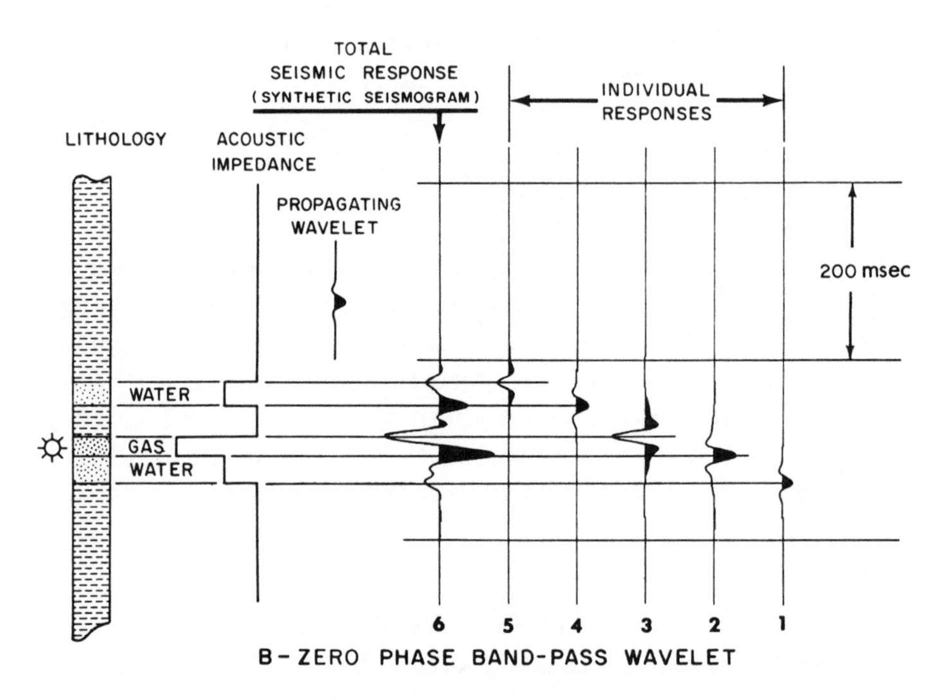

Figure 7.2. Construction of a synthetic seismogram for a sequence of reflecting interfaces. Each interface produces a reflection proportional in magnitude and sign to the reflection coefficient, and the seismic trace is the superposition of these. Diagram assumes a zero-phase wavelet. (From Schramm, Dedman, and Lindsey, 1977; reprinted by permission of The American Association of Petroleum Geologists)

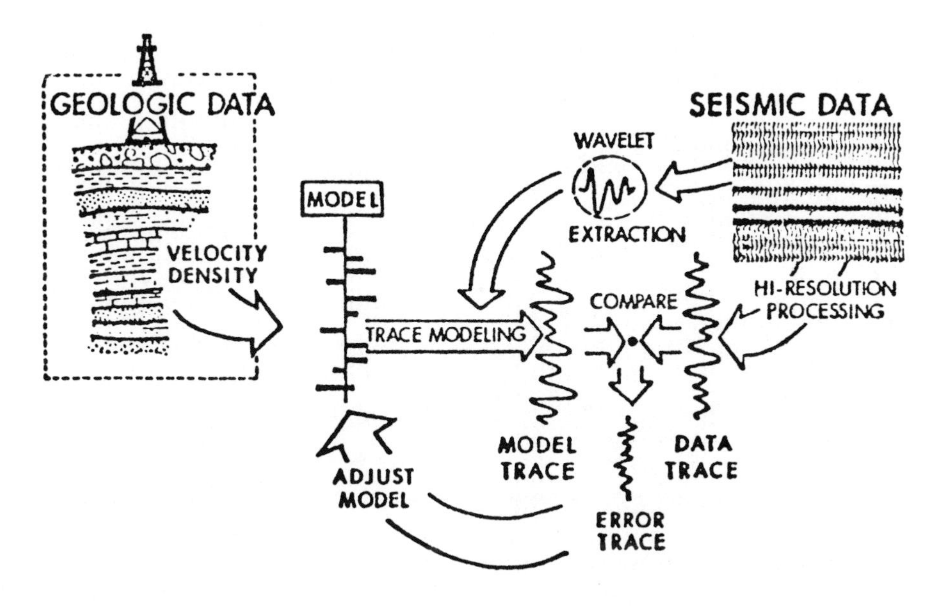

Figure 7.3. Diagram of synthetic seismogram procedure. (From Indonesian Petroleum Association Proceedings, 1978; reprinted by permission of Stommel and Graul)

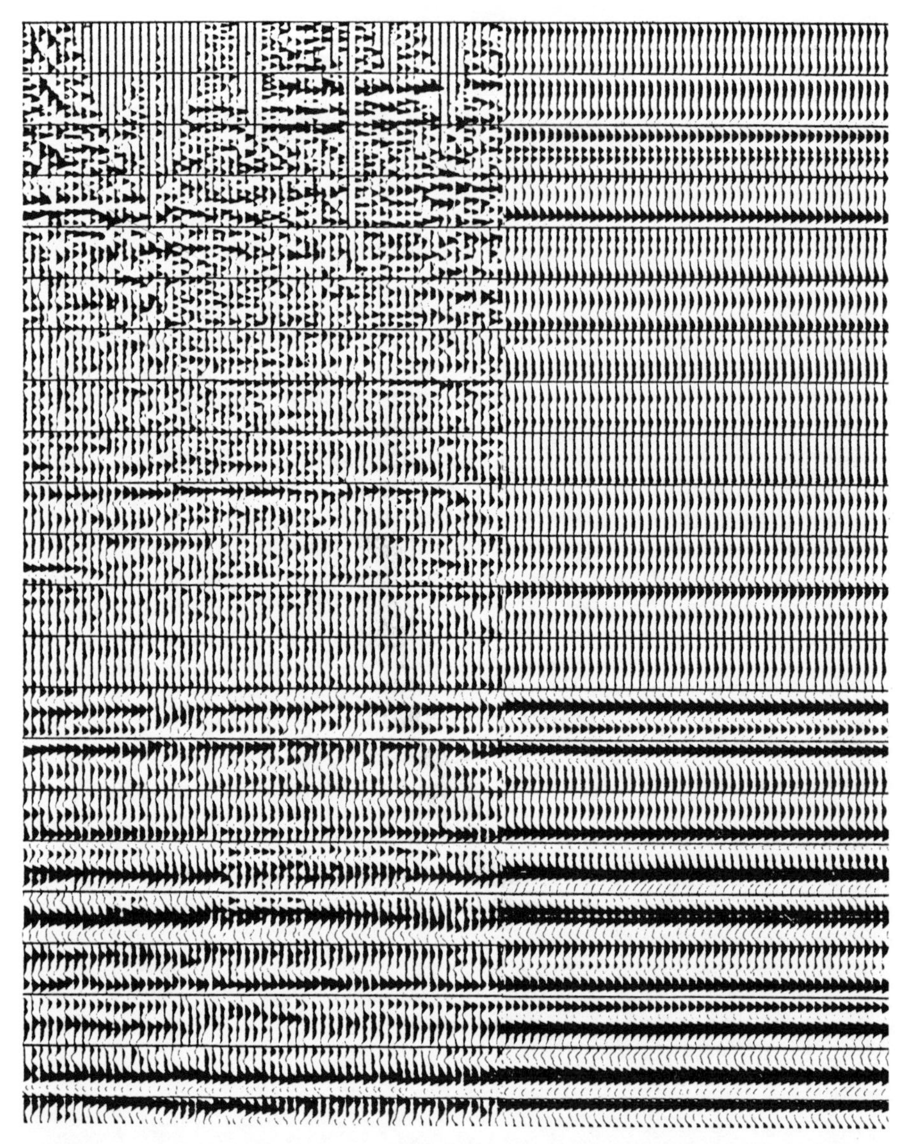

Figure 7.4. Synthetic compared to actual seismograms. Left half of section is a portion of a seismic line, with a well located at the end of the line; right half is a synthetic seismogram made by convolving reflectivity derived from the sonic log in the well with the wavelets derived from the individual seismic traces. (Reprinted by permission of Seiscom Delta)

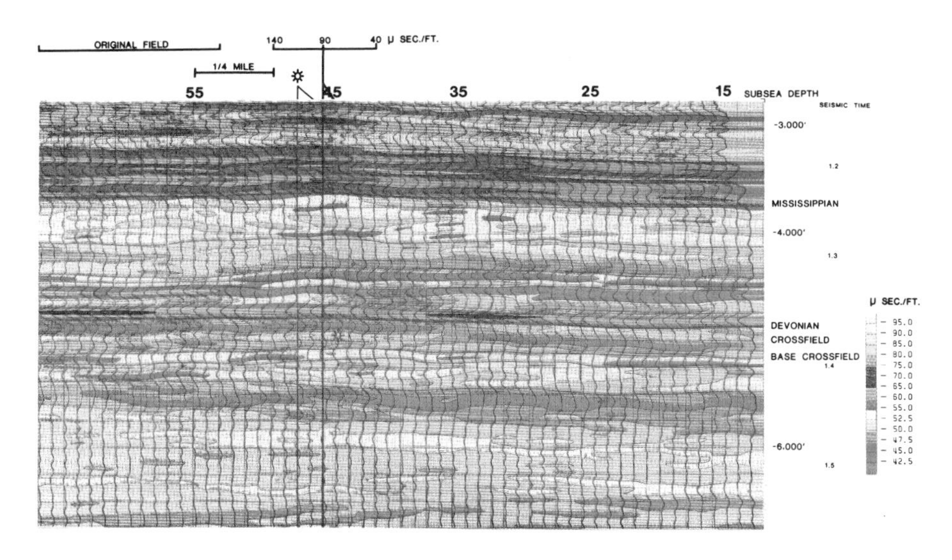

Color plate 1. Seismic log section showing indications of carbonate porosity. An oilfield to the left can be extended to about SP 39; the lower velocity is indicated by the gas symbol on the section. (From Lindseth, 1979; reprinted by permission of The Society of Exploration Geophysicists)

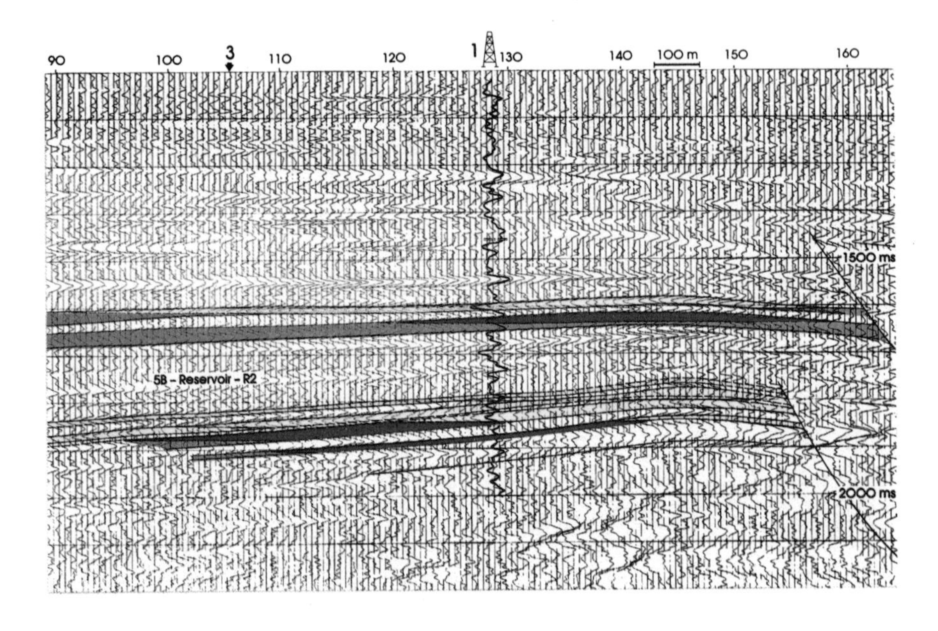

Color plate 2. Relative acoustic impedance display and interpreted productive zones. (Reprinted by permission of Compagnie Generale de Geophysique)

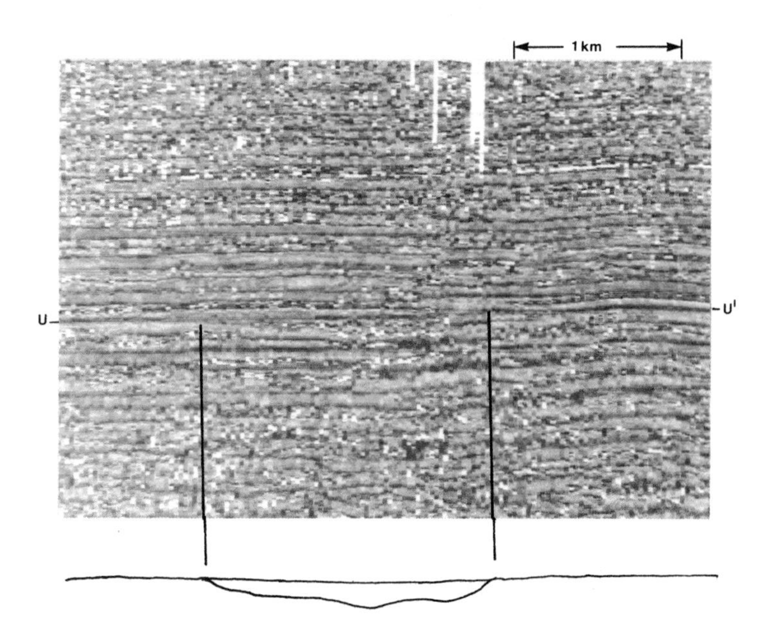

Color plate 3. Line in Williston Basin. UU′ is an unconformity, and an ancient channel is sketched below. Instantaneous frequency has been color coded. (Courtesy Seiscom Delta)

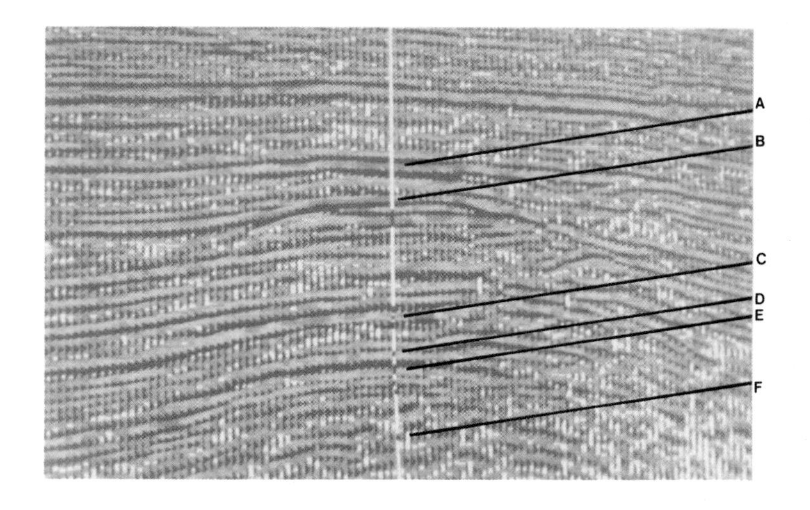

Color plate 4. Productive structure offshore Africa. A is a noncommercial gas zone, B a major gas field (note flat spot), and C, D, E, F are productive zones. The color is amplitude of the seismic envelope. (Reprinted by permission of Seiscom Delta)

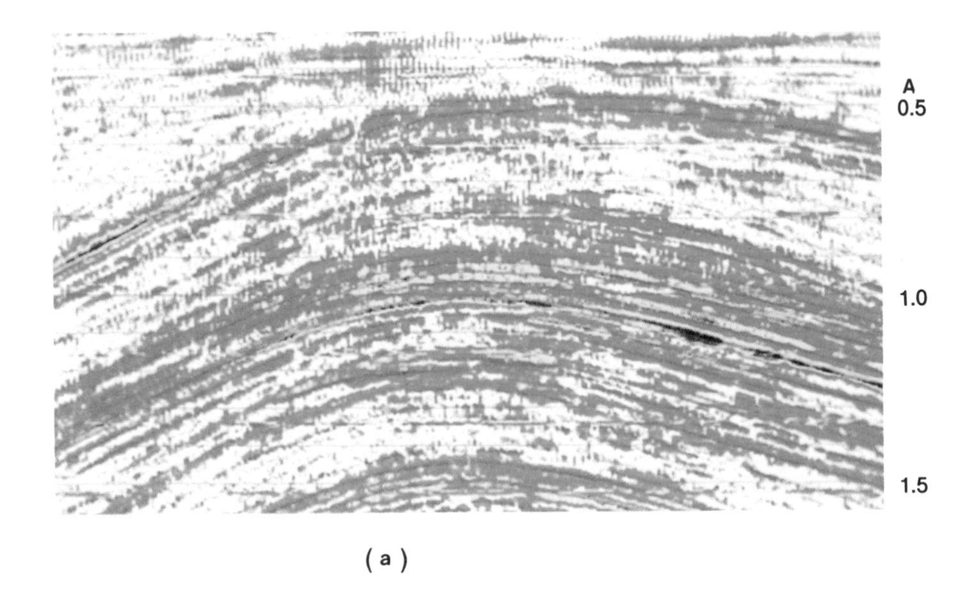

(a)

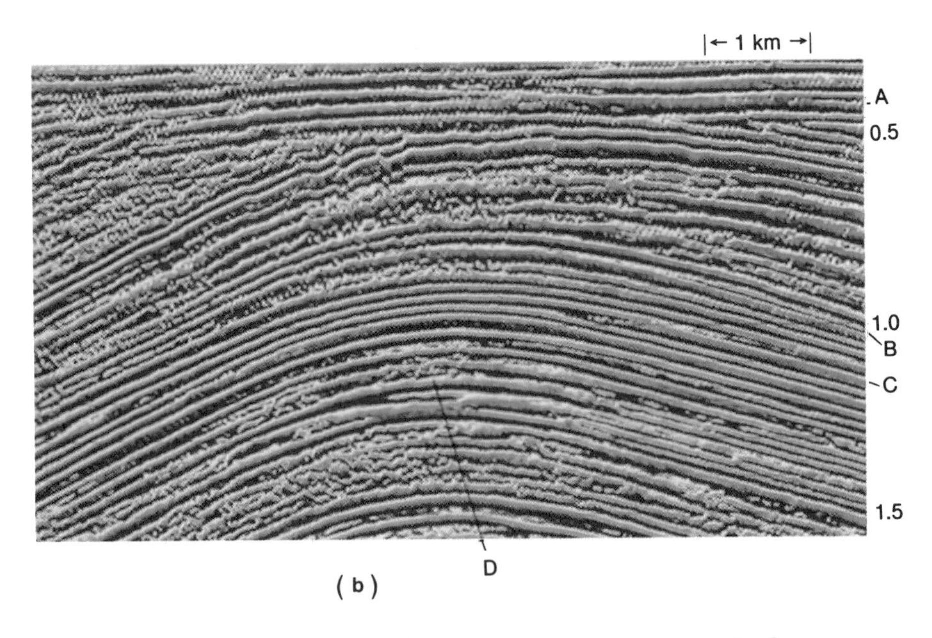

(b)

Color plate 5. Displays of attribute measurements. A, B, C: Angular unconformities; D: flat spot. (a) Display of reflection strength (amplitude); (b) display of phase; (c) display of instantaneous frequency; and (d) display of apparent polarity. (After Tanner, Koehler, and Sheriff, 1979; reprinted by permission of The Society of Exploration Geophysicists)

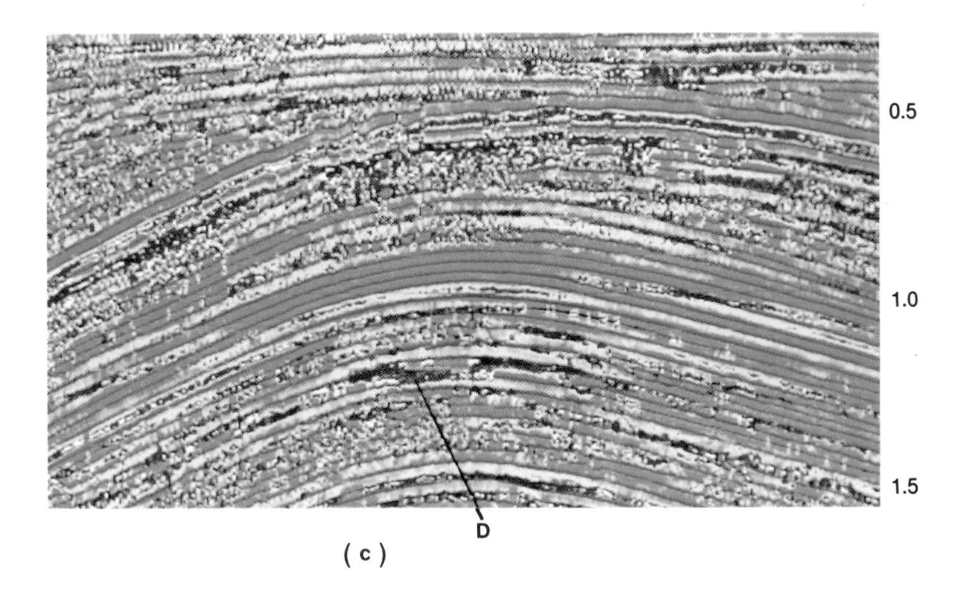

(c)

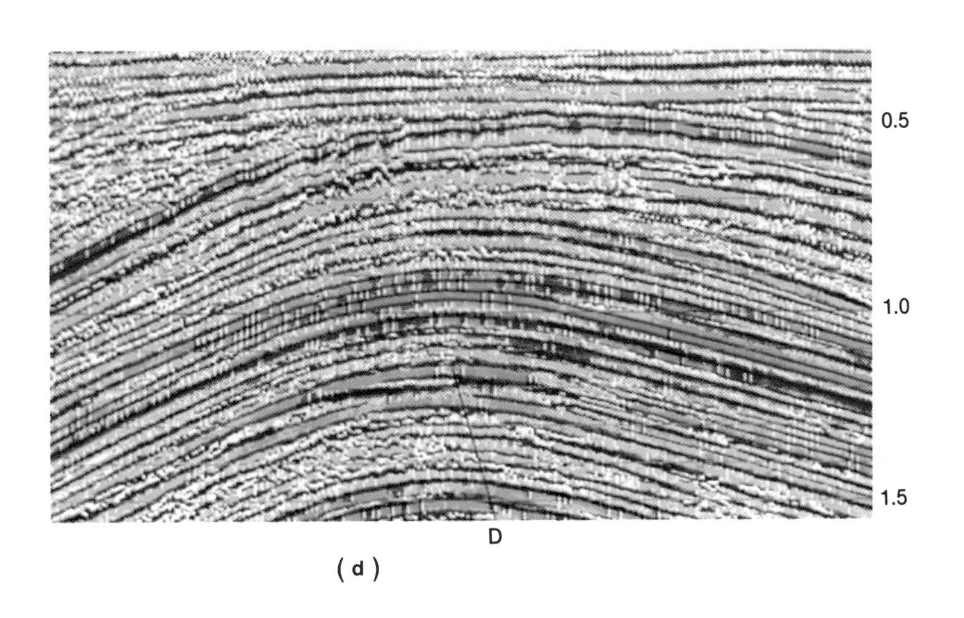

(d)

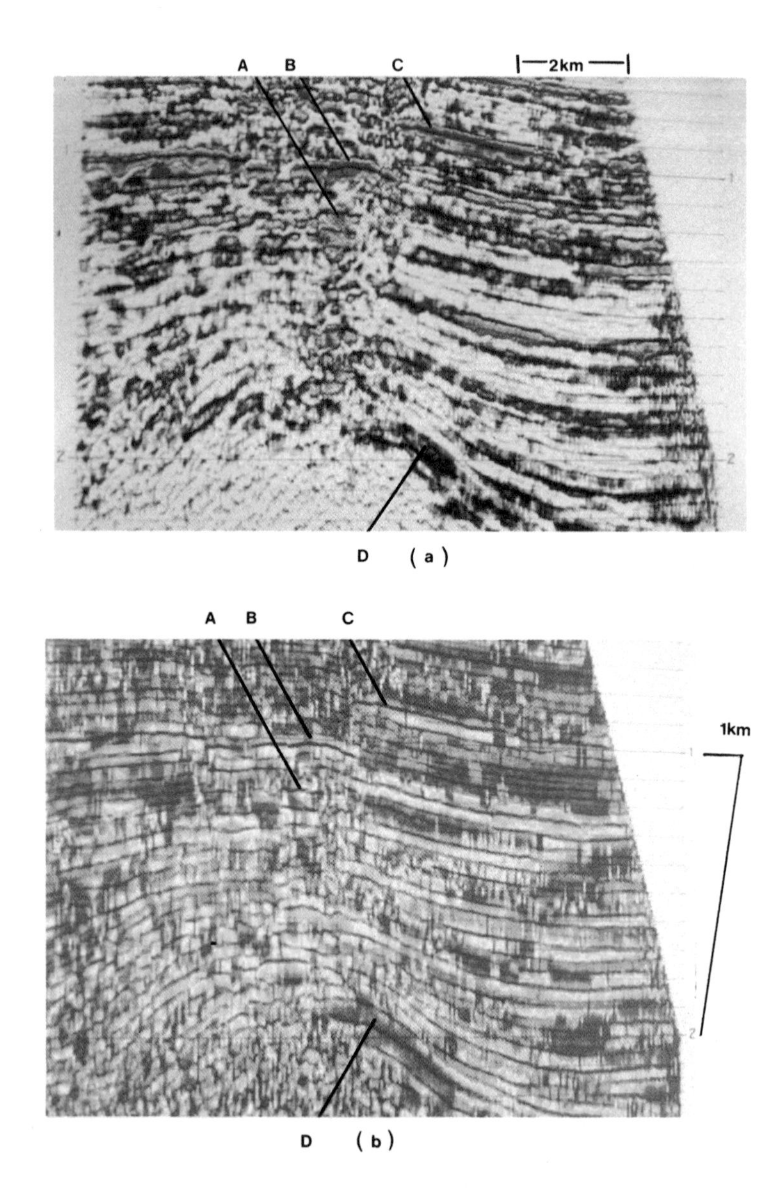

Color plate 6. Line in Gulf of Mexico. A, B, C are gas accumulations; D is a major condensate field. (a) Amplitude of seismic envelope and (b) instantaneous frequency. (Reprinted by permission of Seiscom Delta)

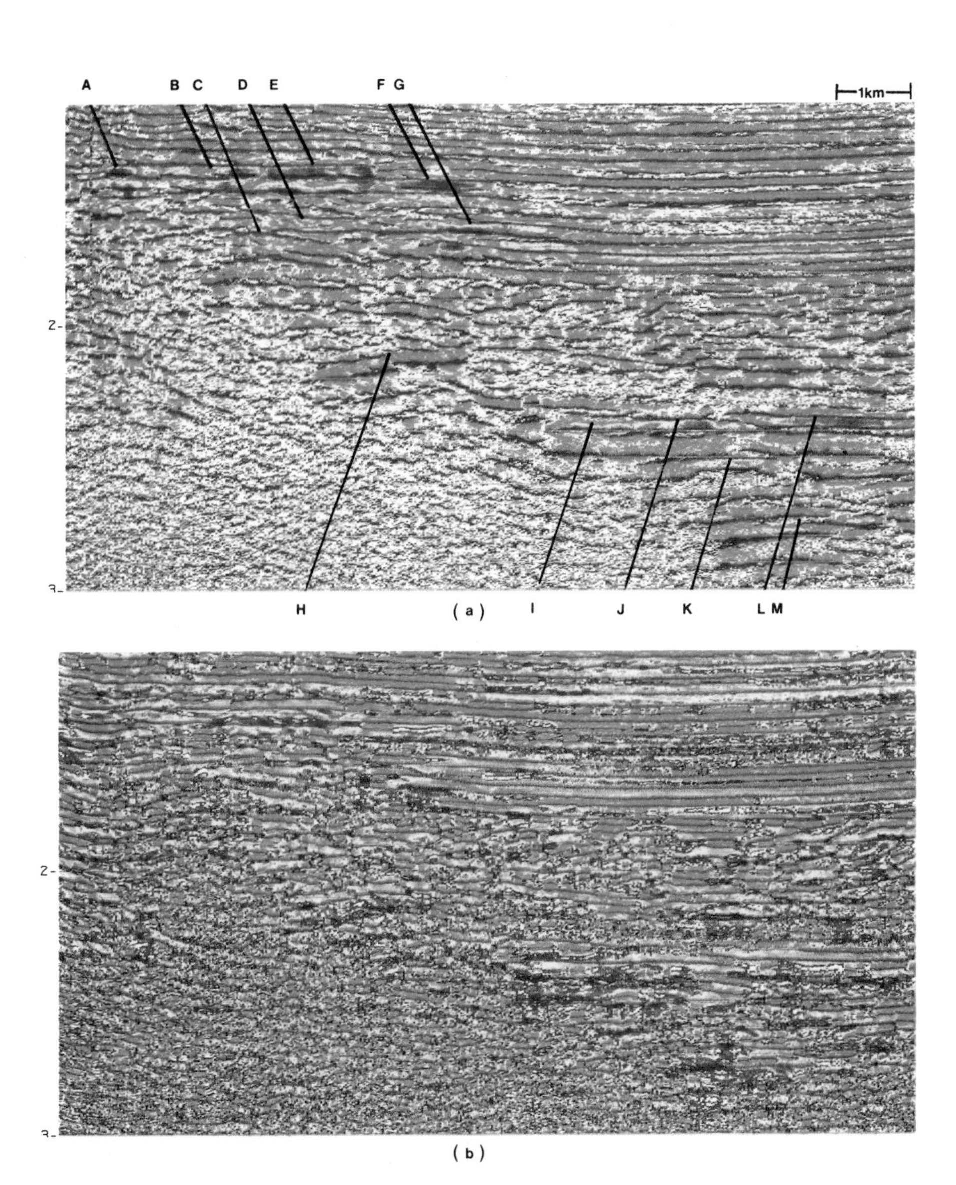

Color plate 7. Productive zone in Gulf of Mexico. Productive thicknesses in meters are A: 2, 8, 4 (3 zones); B: 5, 4, 8; C: 8; D: 11; E: 11, 4, 11; F: 4; G: not drilled; H: 12; I: 18; J: 18, 26; K: 12; L: 18; and M: 7. (a) Amplitude of seismic envelope and (b) instantaneous frequency. (Reprinted by permission of Seiscom Delta)

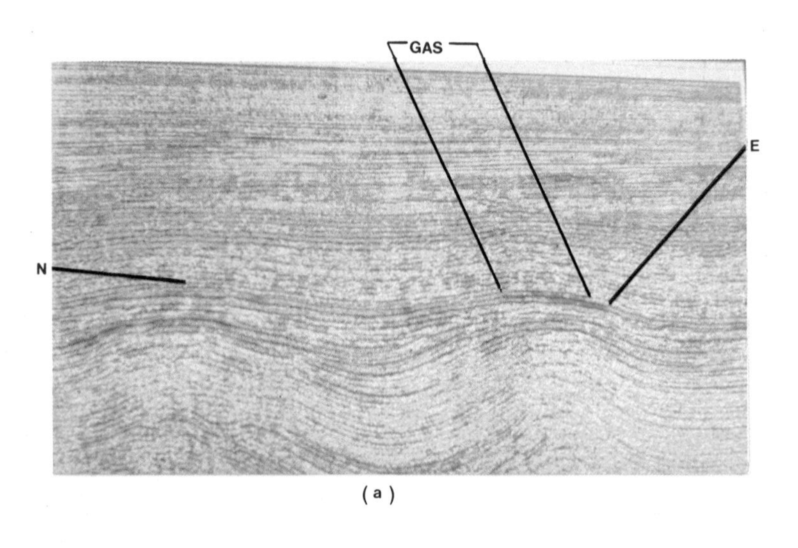

(a)

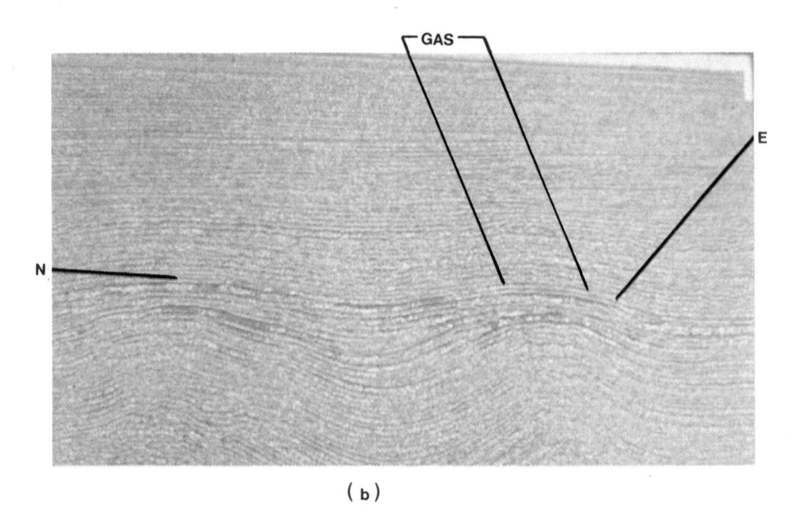

(b)

Color plate 8. Area of production from carbonate rocks. The limits of gas production are indicated; E is the right edge of the oil production, which extends to the spill point on the left; N is a nonproductive equivalent of the gas production zone. (a) Amplitude of envelope of seismic trace and (b) instantaneous frequency. (Reprinted by permission of Seiscom Delta)

often repeatedly in an iterative fashion, until we judge the degree of match acceptable.

Figure 7.4 shows an actual synthetic seismogram (to the right) that was manufactured by convolving the reflectivity determined from a sonic log in a well with the wavelet shapes extracted from actual seismic traces. The left side of figure 7.4 is a portion of the seismic line near the well.

Agreement between the actual and synthetic seismograms is very good. We do not expect a perfect match because the actual and synthetic differ in various regards, such as the effects of density variations, noise contributions, and the fact that they may not be looking at exactly the same subsurface volumes. When a good match can be made, the geologic model can then be varied to see what effects changes in the geology would produce on seismic records. This is done to help identify observed changes on the record with possible changes in the stratigraphic sequence. For example, in figure 7.5, a sand body inserted into the middle of the thick shale member changes the seismic waveshape. If we observe the changed waveshape, we consider a sand body as one of the possibilities.

Conceptually the procedure is simple and straight-forward, but actually the various interfaces that contribute are so close together, and their effects overlap so much (figure 7.6) that it is usually difficult to associate a seismic event with any particular interface.

The problems in actual synthetic seismogram manufacture mainly result from incomplete and inaccurate data. Density logs are usually not available (density is often inferred from the velocity; see figure 6.2). Usually only velocity (sonic) changes are taken into consideration and usually the sonic logs are not complete because portions of the borehole, especially the shallow part, were not logged. The penetra-

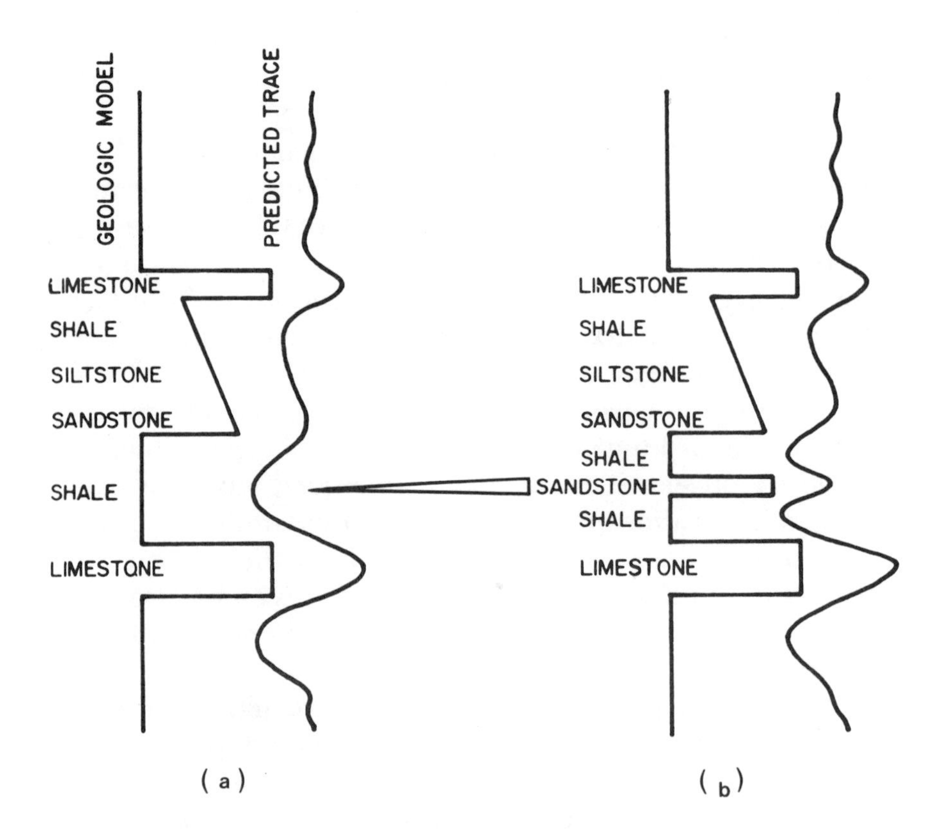

Figure 7.5. Use of synthetic seismograms. If a predicted trace fits actual observations adequately, then a variation of the model suggests the changes in the trace waveform that might evidence such changes. (a) Model and corresponding synthetic seismogram. (b) Model changed to have a sand developed in the middle of the massive shale. (After Clement, 1977; reprinted by permission of The American Association of Petroleum Geologists)

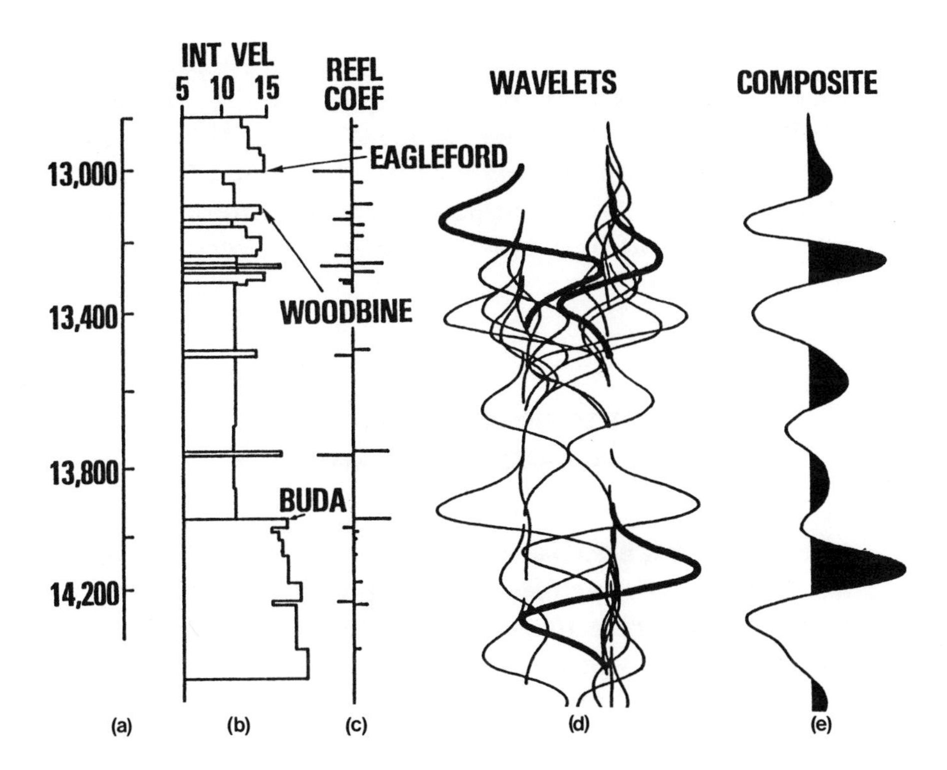

Figure 7.6. Synthetic seismogram construction: (a) depth values are shown, but vertical scale is linear with time; (b) graph of velocity with several formation tops indicated; (c) reflection coefficients calculated by ignoring density; (d) individual wavelets from each of the reflection generators; and (e) sum of all the individual wavelets. (After Vail, Todd, and Sangree, 1977; reprinted by permission of The American Association of Petroleum Geologists)

tion of sonic logs into the formation is not always adequate to represent the unaltered formations reliably; we see disagreements between sonic logs involving different spacings and hence different penetration. Reflections involve a large portion of the subsurface (the Fresnel zone; see chapter 6) and the borehole may not adequately represent the subsurface which affects the seismic trace. Multiples and other types of noise are different on synthetic and actual data. Despite all their limitations, synthetic seismograms are important tools in stratigraphic (and hydrocarbon location) interpretation.

Inverse Modeling: Seismic Log Manufacture

Inverse modeling is also based on the convolutional model. It assumes (figure 7.1b) a noise-free seismic trace to start with and attempts to derive the earth's reflectivity from it, then acoustic-impedance values, and finally the geology. The process is called *inversion.*

Inverse modeling assumes that the amplitude of a seismic trace is proportional to the reflection coefficient and solves the reflectivity equation for the acoustic impedance, or, by assuming the density, solves for the velocity:

$$\begin{aligned}\text{change in logarithm of acoustic impedance} &= 2(\text{reflection coefficient})\\ &= k(\text{seismic amplitude})\end{aligned}$$

$$\text{logarithm of acoustic impedance} = k(\text{integral of seismic amplitude}) + \text{constant}_1$$

$$\text{acoustic impedance} = \text{constant}_2(\text{exponential of } k[\text{integral of seismic amplitude}])$$

Expressing these in the more customary form with mathematical symbols:

$$\Delta(\log \varrho V) = 2R = kA;$$

$$\log \varrho V = k\int A \pm C;$$

$$\varrho V = \varrho_0 V_0 e^{k\int A}$$

where $\varrho_0 V_0 = e^C$; k is called the "scaler" and $\varrho_0 V_0$ the "low-frequency component."

Two "constants" are involved in the solution of these equations. One is the scaler k, which represents the proportionality between amplitude and reflection coefficient. It is implicitly assumed that all amplitude variations result from reflectivity changes and that the amplitude information has been preserved faithfully in the processing. The second "constant", $\varrho_0 V_0$, results from the constant of integration. It is implicitly assumed that all required information is available, including both high- and low-frequency (and d.c.) components. The frequency content of seismic data is severely band-limited compared to that of sonic logs (figure 7.7). We usually recognize that well logs involve much higher frequencies than seismic data and do not expect to recover the high-frequency detail. Seismic data are also missing the low frequencies and it is these involved in $\varrho_0 V_0$. In the foregoing we have put "constants" in quotation marks since neither k nor $\varrho_0 V_0$ is actually constant.

Neither constant can be obtained from the seismic data by the inversion process. Commonly the constants are evaluated from either sonic logs or stacking velocity data. Even where nearby wells have been logged, usually only velocity measurements are available so that the solution assumes the

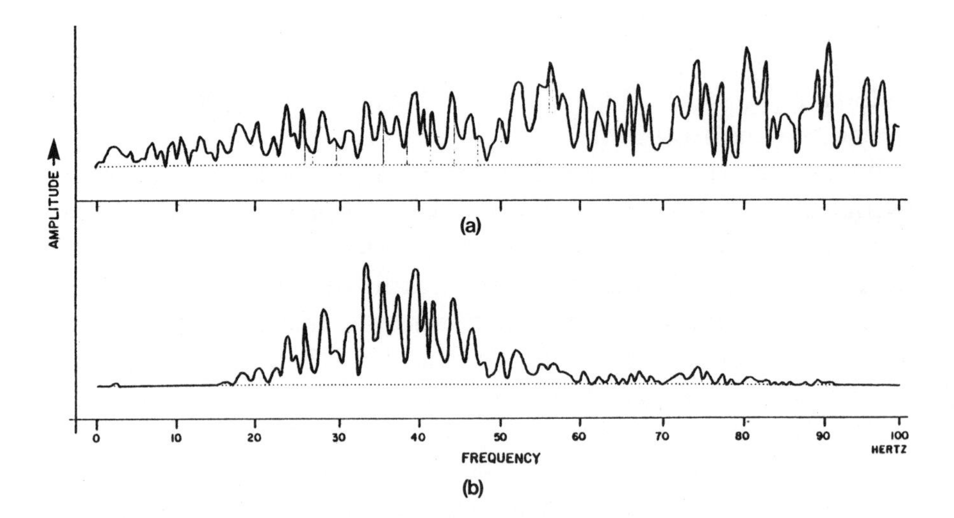

Figure 7.7. Amplitude spectra of well log and seismic trace: (a) spectrum of reflection coefficients and (b) spectrum of a seismic trace. (From R. O. Lindseth)

density. The usual assumptions are either that the density is constant or that it has a functional relationship to the velocity, the most common relationship being Gardner's (see figure 6.2), that the density is proportional to the 1/4 power of the velocity.

The result of inversion is usually called a *seismic log*, a *synthetic acoustic impedance log (SAILE trace)*, a *simulated* or *synthetic sonic log*, or an *instantaneous velocity log*. The justification for calling the result "velocity" is that the constraints applied are usually velocity so the result has the dimensions of velocity.

The problems with trace inversion are mainly caused by the inability to separate completely reflectivity effects from noise. Trace inversion is mainly used to study trace-to-trace variations, and the determination of the constants in the inversion process does not directly affect such variations, although they alter their apparent magnitude.

Inversion assumes a noise-free input, so it must be almost the last step in data processing since all the noise-removal processing must precede it. Since noise effects cannot be removed completely residual noise is a basic limitation. The separation of wavelet from reflectivity effects must precede inversion, and wavelet processing designed to accomplish this has been discussed somewhat in chapter 2. However, the separation of wavelet and reflectivity is not unique and also remains a limitation to inversion.

Inversion is essentially an operation on amplitude measurements, and amplitude depends on many factors besides reflectivity (figure 7.8). Adequate handling of amplitude-distorting factors is often the key to good inversion results. Migration (see chapter 2) is an important factor in remedying some types of amplitude distortion and thus properly precedes inversion.

Examples of Inversion

Figure 7.9 shows a seismic record that has been inverted to seismic logs (instantaneous velocity). Each seismic log trace results from the inversion of one seismic trace. A sonic log is plotted (heavier, in the center) to the same scale for comparison. In some portions the copy is good, in others poor. We do not expect the two to be identical because they involve a number of different assumptions and are made from basically different sources of information. In some places the agreement is better than first apparent; for example, at A the seismic data show a simple peak and the sonic log a double peak, but if the event is followed on the seimic logs it becomes a double peak also. The sonic log of figure 7.9 wanders somewhat across the center gap; this is because the sonic log contains low-frequency information not available from the seismic data, as was shown in figure 7.7. Often sonic logs are filtered to the seismic passband before making comparisons but low-frequency filtering was not used on the sonic log plotted on figure 7.9.

Color plate 1 shows an inverted seismic line where the color-coded interpreted velocity is displayed. A porous zone in the carbonates provides a gas reservoir that extends partway in from the left end of the line. The line shows a region of lower velocity indicated by the gas symbol that turned out to be a field extension. The limit of porosity development is considered to be at SP 39.

Color plate 2 shows an inverted section with the velocity of reservoir units color-coded, using yellow to indicate gas-bearing, red for oil-bearing and blue for water-bearing. Color is used in two ways that are sometimes mixed up: (1) to display the data in a way that is easier for an interpreter to comprehend, and (2) to display an interpretation. One sometimes combines the two as was done in color plate 1, where the blue colors were assigned to the velocities that were

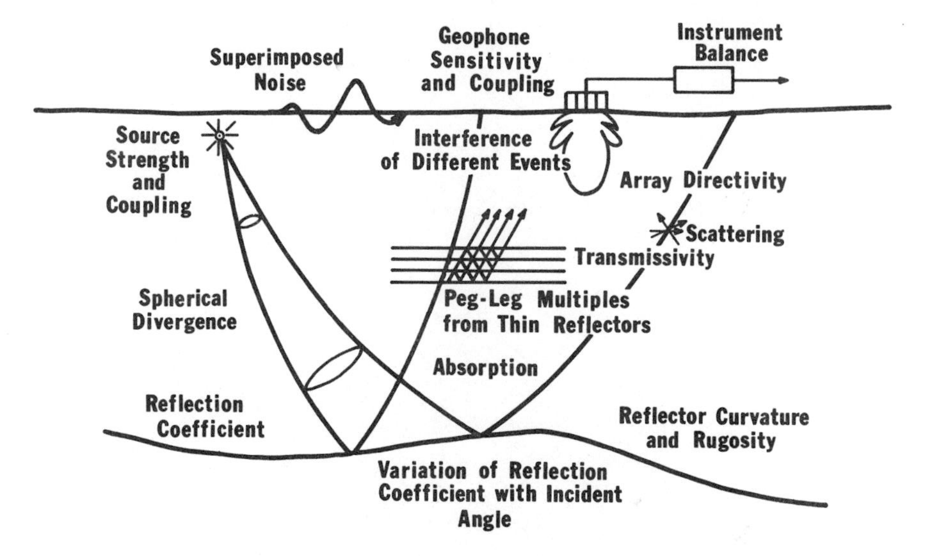

Figure 7.8. Factors that affect seismic amplitude. To see the variations because of reflection coefficient changes, variations because of the other factors have to be compensated for. (From Sheriff, 1975; reprinted by permission of The European Society of Exploration Geophysicists)

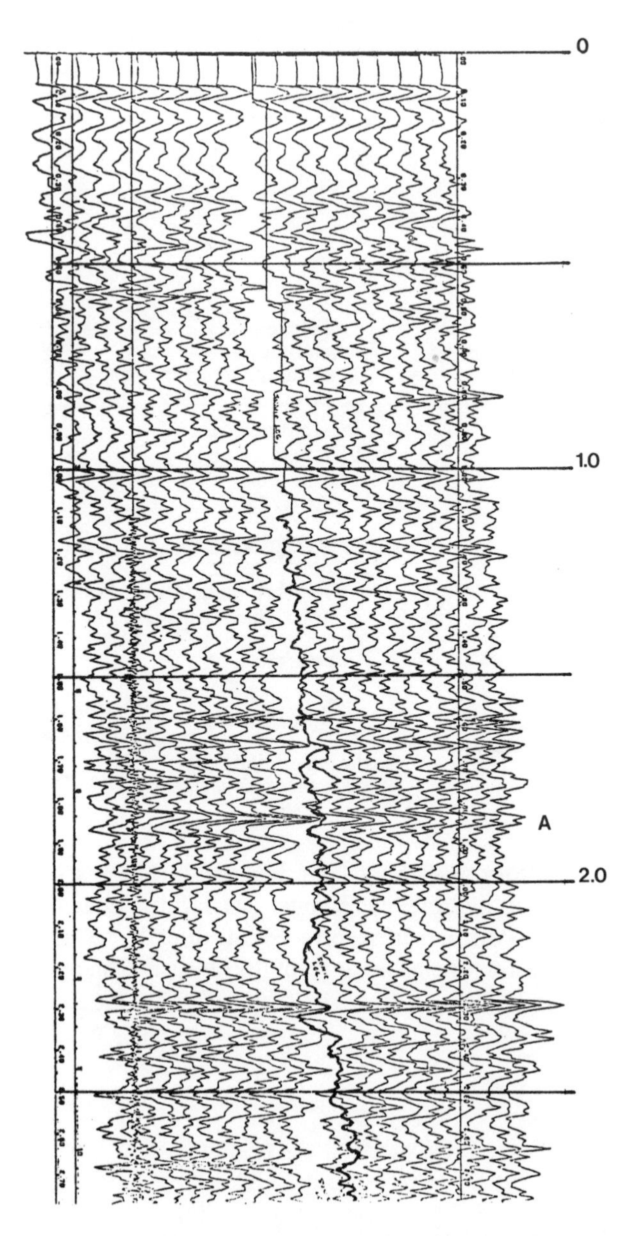

Figure 7.9. "Instantaneous velocity" calculated by inversion of seismic traces. The heavy curve in the center is data from an actual sonic log. The horizontal velocity scale is such that the trace spacing represents 380 m/s. (Reprinted by permission of Seiscom Delta)

interpreted to be carbonates, the yellows and reds to those velocities interpreted to be sands, and the greens to those thought to be shales. The juxtaposition of contrasting colors (for example, red to green in color plate 1) tremendously affects the features that most impress those who look at the section. The choice of colors can thus prejudice an interpretation.

8
Reflection Character Analysis

Reflection character analysis involves study of the changes in the waveshape of one or more reflections associated with a reflectivity change. The objective of reflection character analysis is to interpret trace-to-trace waveshape changes in terms of changes in the stratigraphy or fluid in the pore spaces. Reflection character analysis usually involves both forward and inverse modeling (see chapter 7).

The initial focus of reflection character analysis was on hydrocarbon detection. The properties of a reservoir generally change with the nature of the interstitial fluid, changing the contrast with respect to the reservoir seal and, consequently, the reflection waveshape. The features of the reflection waveshape of special interest are timing, amplitude, polarity, and frequency.

Simple Situations

Figure 8.1 shows the waveshape expected from several types of simple contrasts apt to be encountered in the earth. The reflection from a simple change in acoustic impedance (a step) has the same waveshape as the downgoing wavelet, the reflection amplitude is proportional to the size, and the polarity to the direction, of the change. For a very thin bed, the reflection waveshape is the derivative of the downgoing waveshape. Where the reflecting bed is a 1/4 wavelength thick, the reflections from the top and base interfere constructively and give exceptionally large amplitude, an effect called tuning. Where the bed is very thick, the reflections from the top and base are completely separate. Where the acoustic-impedance change is gradational, the reflection waveform is the integral of the downgoing waveform.

Figure 8.2 shows reflections from a thin bed where the acoustic impedance is gradational. The gradational base might represent a regressive sand with a gradual upward increase in sand content.

Much modeling is concerned with structural features or with combinations of structural and stratigraphic features. Figure 8.3 shows the reflections from a simple syncline; changes occur in both amplitude and waveshape, so that not all waveshape changes imply changes in stratigraphy. Models are often much more complicated and can involve lithology changes and hydrocarbon traps as well as structural features (see figure 8.4). A use of modeling is learning how effects appear and whether they are large enough to be seen in actual seismic data. While models may be quite complicated, they are still very simple compared with the real earth.

Amplitude as an Important Discriminant

The model of figure 8.5 concerns a sand body embedded within a shale. The model shows variations in thickness, in depth, and in whether the sand is a single unit or two. The individual members are so thin and close together that they are not individually resolved. The waveshape changes but little and the amplitude carries the information as to the amount of sand in the interval where the sand is thin; the amplitude is nearly constant where the gross thickness is greater than a 1/4 wavelength. Whether the sand is in one or two bodies makes little difference as long as the sands are thin and sufficiently close together as to be individually unresolvable.

The use of amplitude as a measure of the thickness of a thin bed is theoretically sound but often difficult to apply for want of an amplitude reference. Figure 8.6 is similar to what one would expect for an enlargement of the wedge end of figure 6.9 (although made independently). It shows the build up in amplitude at the 1/4 wavelength thickness and the decrease in amplitude as the wedge thins. If one had an amplitude reference, that is, if one knew the amplitude to expect for a thick bed or if one could determine where the bed was a 1/4 wavelength thick, then amplitude could be a useful measurement to tell the thickness of the bed when thin.

Figure 8.6b also shows amplitude versus thickness graphs for wedges where the contrasts at top and base of the wedges differ. All have been normalized by setting the maximum peak-to-trough amplitude values for a thick wedge equal to 1.0. The reinforcement at 1/4 wavelength thickness results from the trough of the reflection from the base of the wedge being superimposed on the peak of the reflection from the top of the wedge at this thickness; the exact waveform under consideration influences the magnitude of the value at the tuning thickness. Where the reflection coefficients at top and

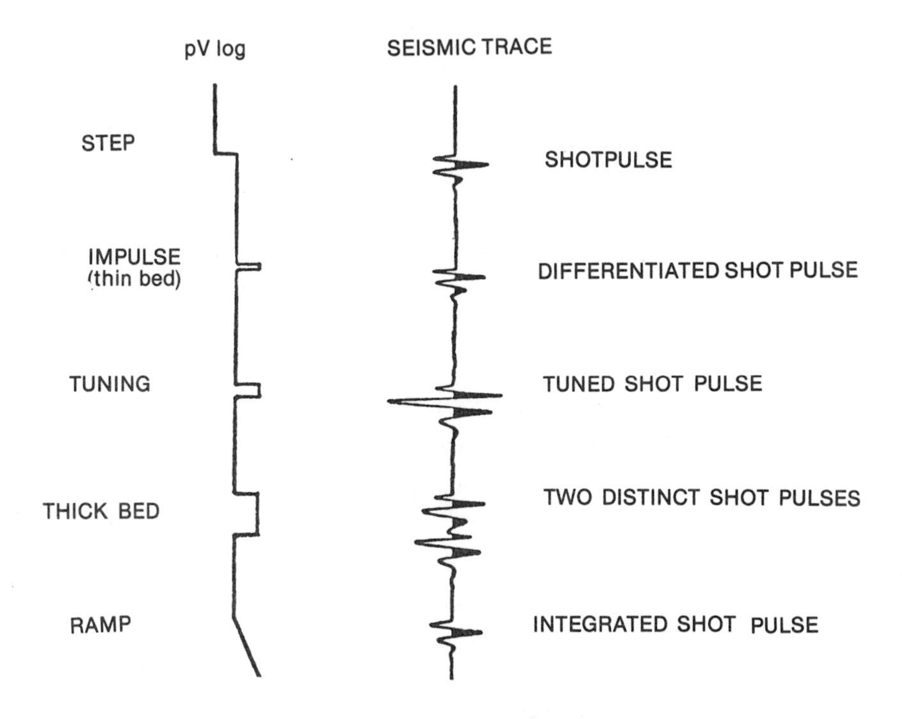

Figure 8.1. Reflections from simple interfaces. (After Sengbush, Lawrence and McDonal, 1961; reprinted by permission of The Society of Exploration Geophysicists)

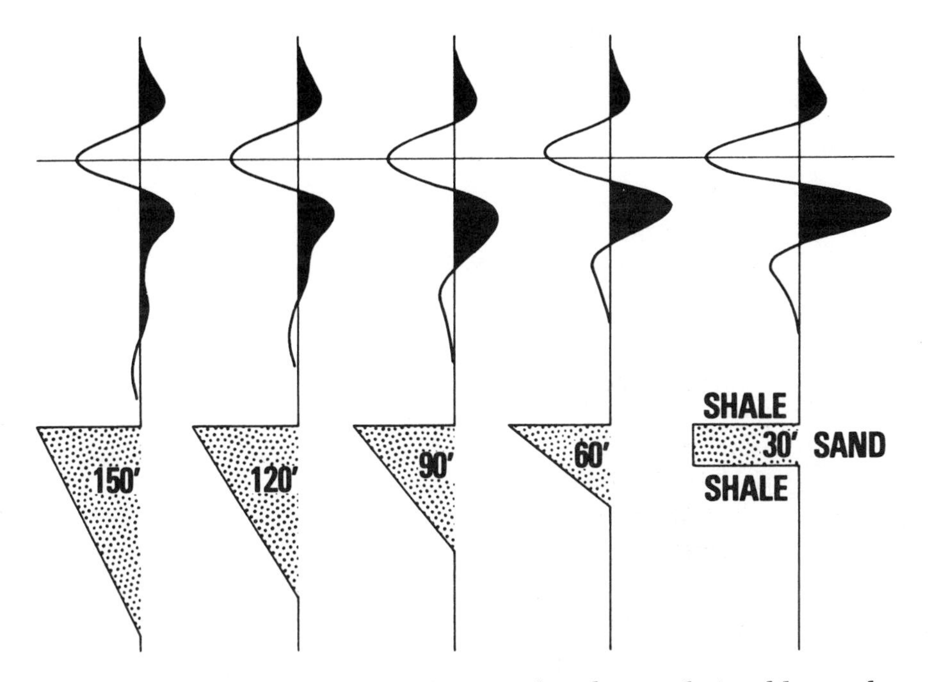

Figure 8.2. Seismic response for a sand with a gradational base. The 30-foot thickness is about 1/8 wavelength. (From Neidell and Poggiagliolmi, 1977; reprinted by permission of The American Association of Petroleum Geologists)

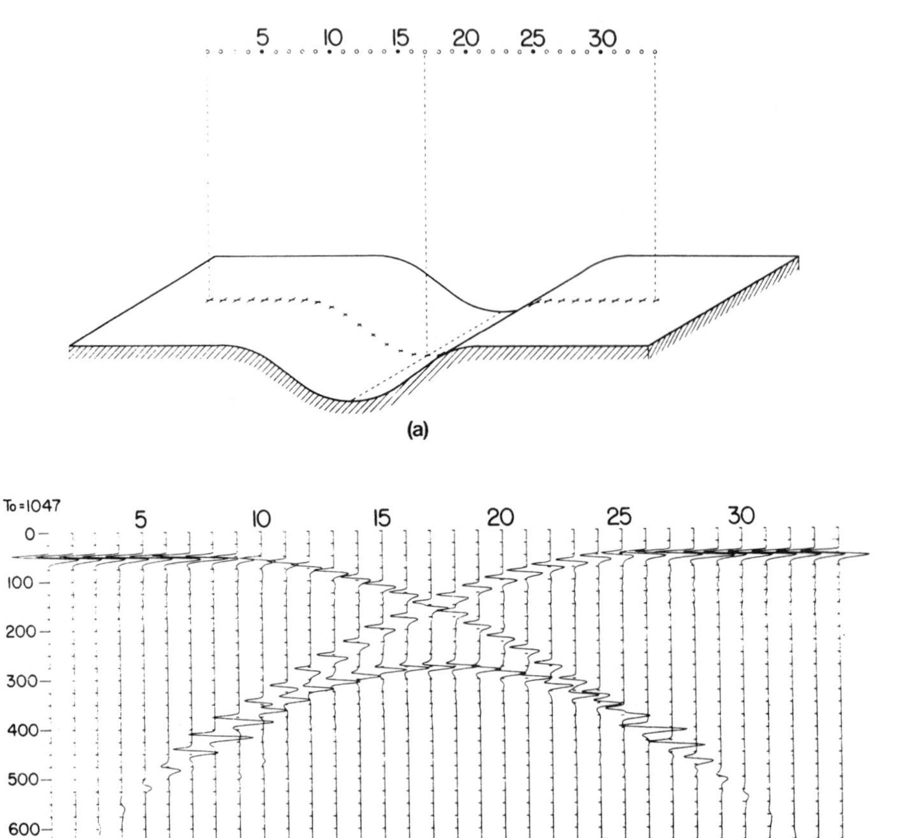

Figure 8.3. Model of a simple syncline: (a) model and (b) reflection section across the model. (From Hilterman, 1977; reprinted by permission of The American Association of Petroleum Geologists)

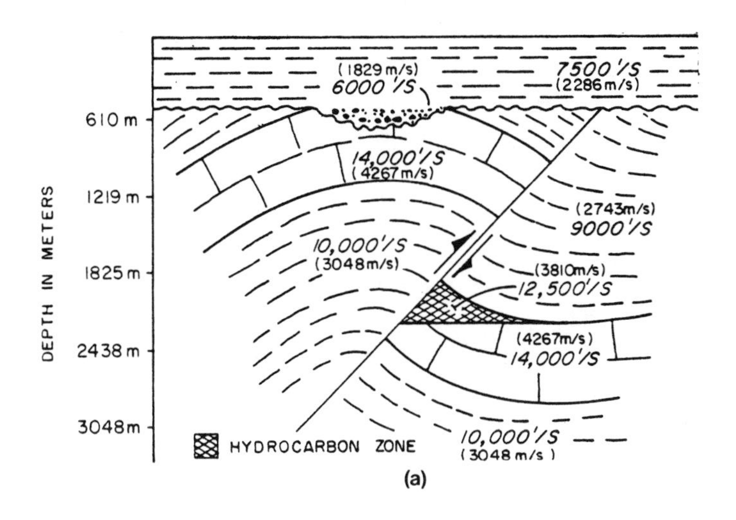

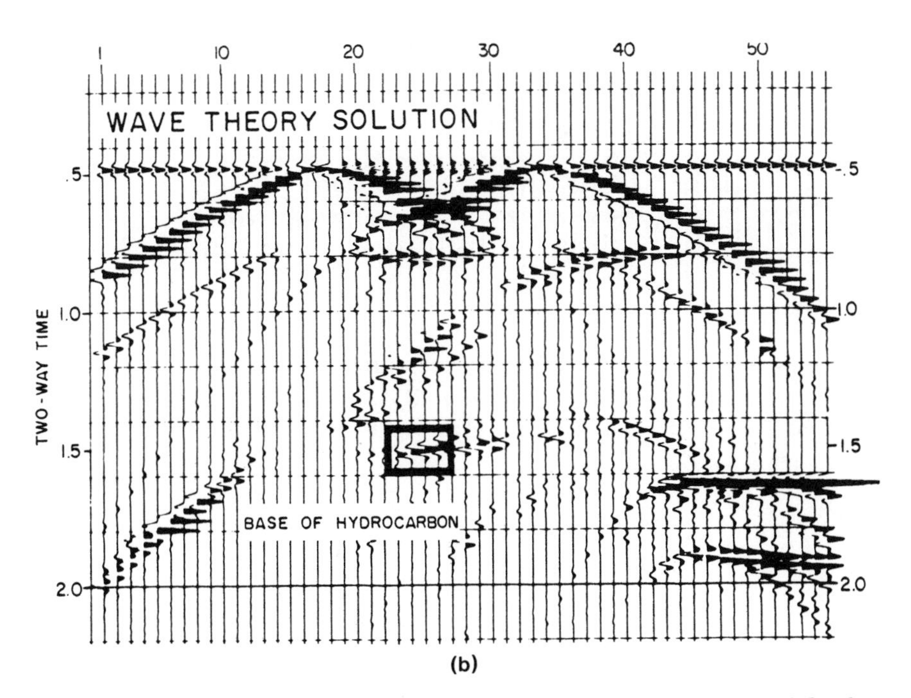

Figure 8.4. Model combining structural, stratigraphic, and hydrocarbon accumulation features: (a) model and (b) synthetic seismogram. (From Neidell and Poggiagliolmi, 1977; reprinted by permission of The American Association of Petroleum Geologists)

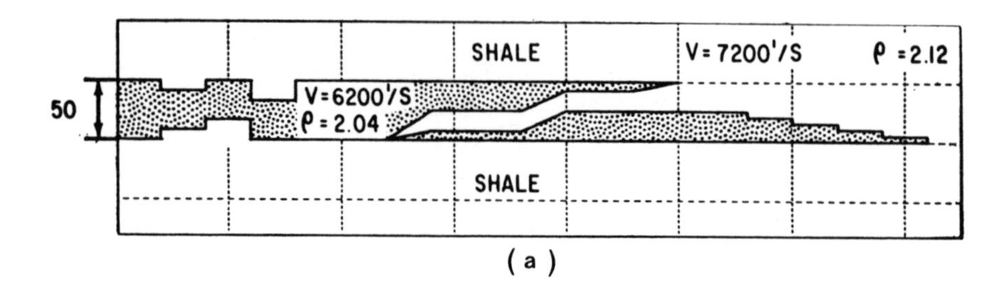

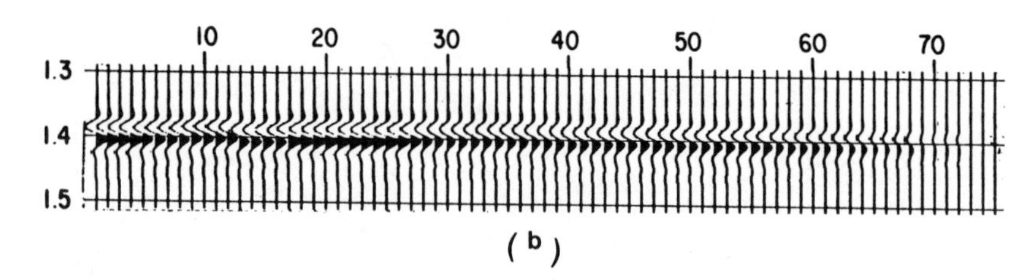

Figure 8.5. Model of sandstone variations, imbedded in shale: (a) model and (b) synthetic seismogram. (From Meckel and Nath, 1977; reprinted by permission of The American Association of Petroleum Geologists)

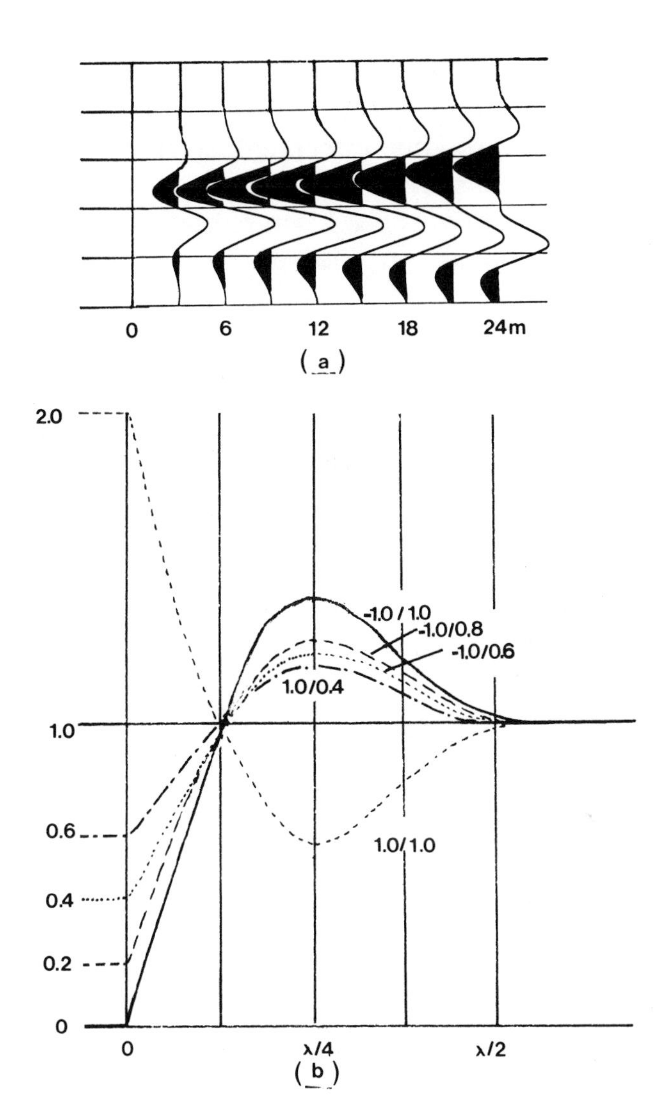

Figure 8.6. Reflections from a wedge imbedded in a constant-velocity medium for a zero-phase wavelet. (a) Reflection waveshape; the wedge thicknesses are given in meters for a velocity of 1,525 m/s. (b) Maximum peak-to-trough amplitudes where the reflectivity at top and base of the wedge has ratios 1.0/1.0, -1.0/1.0, -1.0/0.8, -1.0/0.6, -1.0/0.4. The -1.0/1.0 curve applies to the wedge in (a). Wedge thickness is given in wavelengths. (Figure 6.9 involved a 1.0/1.0 contrast and figure 6.11 a -1.0/1.0 contrast.)

base of the wedge are unequal, there is still a contrast after the wedge has completely pinched-out. This figure also shows the antiresonance effect where the reflections at top and base are of the same sign; in this instance the waveforms tend to a minimum at the 1/4 wavelength thickness. This latter effect was also seen in figure 6.9.

Case Studies

Figure 8.7 is from a case history study of the seismic effects of a pinchout of two sands. A well log in the area under study was modified in several ways thought to be possibilities, and synthetic seismograms were made for each of the possibilities. Figure 8.7 shows considerable change in waveshape between the 0- and 50-feet thick models, but increasing the thickness to 100 feet produces only a small change in amplitude. This might be thought of as a negative result if the objective was to determine the sand thickness within this range. Sometimes the conclusion of a synthetic seismogram study is that we are unable to resolve the reservoir parameters in the range of most concern.

The study of which figure 8.7 is part defined a set of criteria that was useful in surveying the area. These included rules regarding which events provided the best reference datums for study of other events; which events were interference composites whose changes were not significant; and which events varied in amplitude or waveshape so as to indicate pinchouts of the sands it was desired to map.

Figure 8.8 poses the question of whether the manner in which a sand is distributed within a shale bed makes much difference on its seismic appearance. It appears that the distribution does not significantly affect the waveshape if the total interval involved is less than the resolvable limit. The reflection amplitude depends mainly on the total thickness

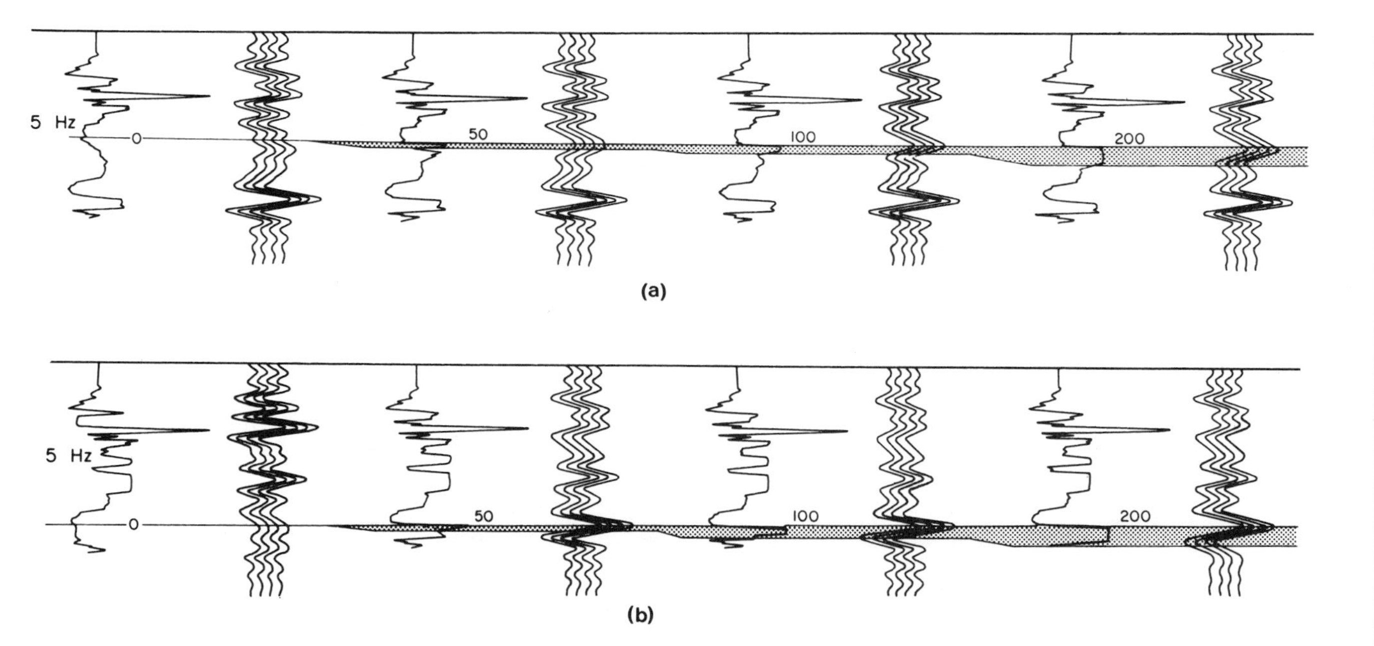

Figure 8.7. Synthetic seismogram study of effects of sand thickness. Diagrams are paired showing a velocity model and a group of four traces of a synthetic seismogram. The models in each row are identical except for the sand thickness. (a) and (b) Variable thickness of sands in two different intervals. (After Galloway, Yancey and Whipple, 1977; reprinted by permission of The American Association of Petroleum Geologists)

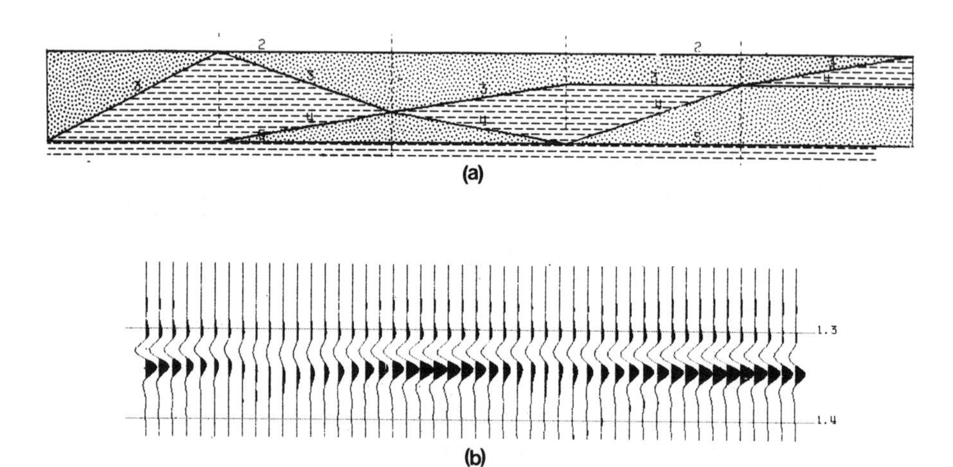

Figure 8.8. Model of sand variations within a massive shale. The amplitude relates to the total sand thickness. (After Meckel and Nath, 1977; reprinted by permission of The American Association of Petroleum Geologists)

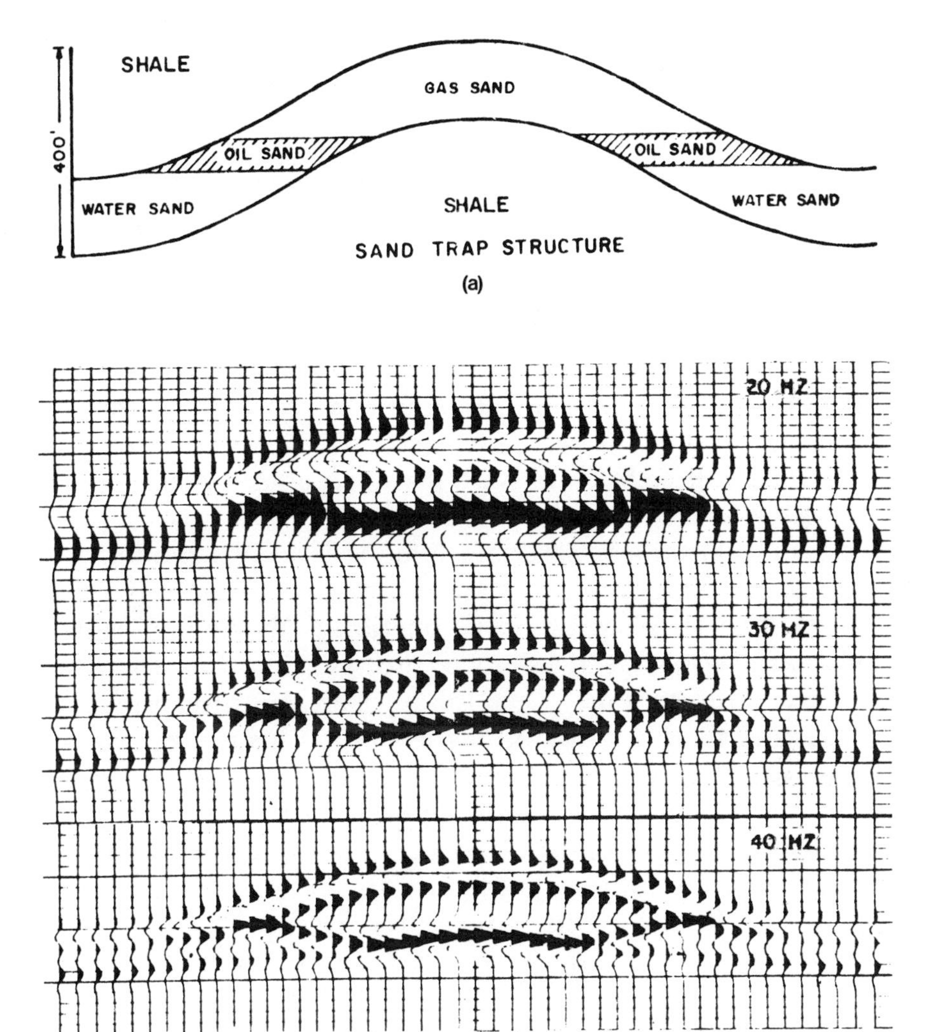

Figure 8.9. Model of accumulation in a simple anticline: (a) model and (b) synthetic seismograms for 20, 30, and 40 Hz Ricker wavelets.

of sand, and whether the sand is in one or several bodies is difficult to ascertain.

A fairly simple anticline containing gas/oil accumulations is modeled in figure 8.9. Wavelets with dominant frequencies of 20, 30, and 40 Hz have been used to make synthetic seismograms. The sand thickness is just about a 1/4 wavelength in thickness in the water-sand case for the 20 Hz wavelet. The reflection from the base of the sand appears at first glance to be much stronger than that from the top of the sand, but this is an illusion produced by the variable-area display, which emphasizes peaks by shading them in, so that the trough associated with the sand top does not make as much impact as the peak from the sand base, although they are equal in amplitude. Often sections are produced both with normal polarity and with inverted polarity so that this bias will not affect the interpretation.

A combination stratigraphic-structural trap is modeled in figure 8.10, a change in permeability providing the updip closure. The termination of the portion with the high reflection coefficients (the gas-sand portion) against the silty sand produces diffractions. This might lead an interpreter to postulate a fault as providing the updip closure; to seismic waves, the geologic mechanism that terminates the reflector is not of concern. The strengthening on the third trace from the right is because of geometrical focusing in the syncline; a weakening of amplitude occurs on the crest of the anticline because of the defocusing curvature.

Figure 8.11 is an attempt at the realistic modeling of ancient river channels. The two sides of the model were designed from maps and sections across two known fields. The channels and channel-sand deposits produce both variations in the amplitude and in the time interval between reflection picks that might be associated with the base of the Mowry and the top of the Dakota formations. These variations might be measured and greater reliability would be placed where

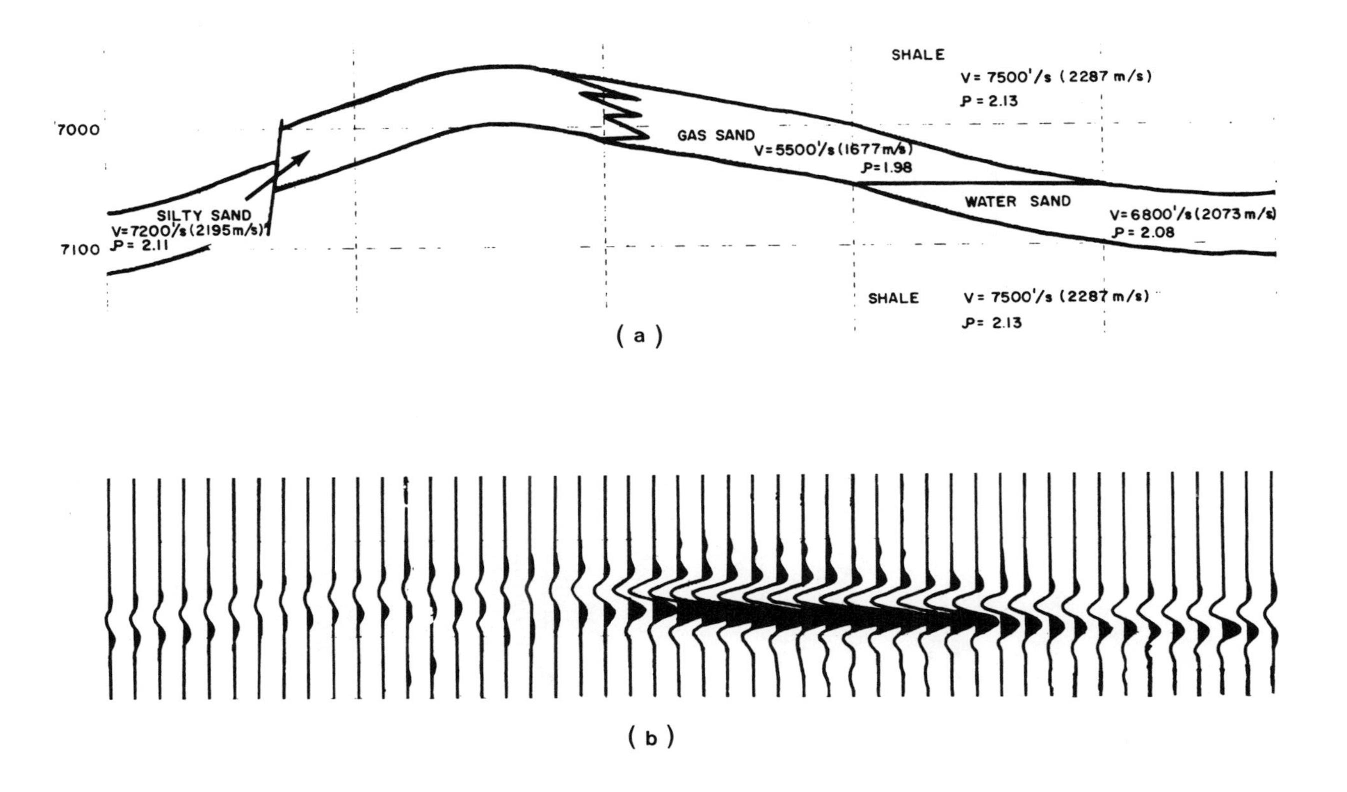

Figure 8.10. Model of a semi-stratigraphic gas accumulation. Updip closure is stratigraphic but a gas-water interface limits the accumulation downdip. (After Neidell and Poggiagliolmi, 1977; reprinted by permission of The American Association of Petroleum Geologists)

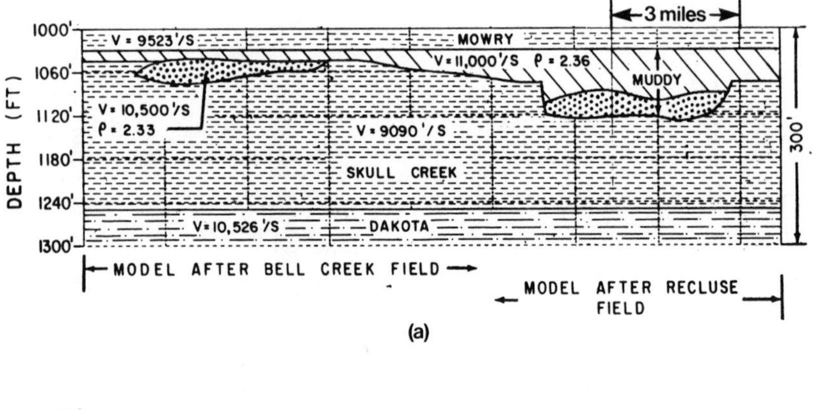

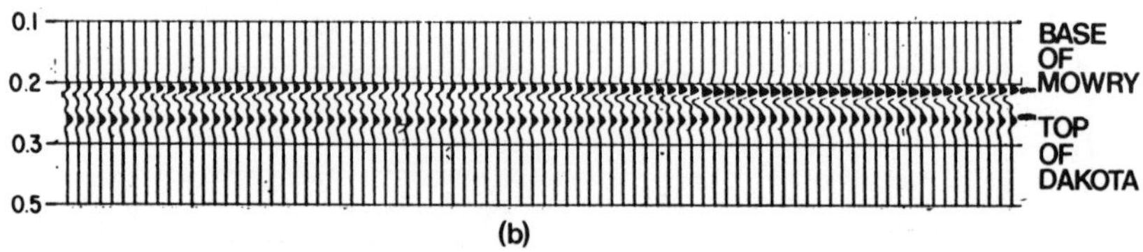

Figure 8.11. Model of stratigraphic traps in Powder River Basin: (a) model and (b) synthetic seismogram. (From Meckel and Nath, 1977; reprinted by permission of The American Association of Petroleum Geologists)

they corroborate each other. Note that the individual Muddy formation interfaces are generally too close together to be resolved. Naming reflections, as has been done here and as is the usual practice, should not lead an interpreter to believe that mapping of the respective cycles gives a map of the particular interface. Reflections are almost always the interference composites of many nearby reflections, a change in any one of which may change the composite. The naming of certain phases after formation contacts provides a convenient way to discuss effects and helps focus attention on the zone of interest despite possible misleading implications. Another channel sand, in a somewhat deeper channel, is shown in color plate 3 where attribute measurement and display have been used to enhance its detectability.

A case history of a search for accumulations in channel sands is described by Clement (1977). He had a number of well logs in the study area (figure 8.12) as the basis for his study. The problem here is one of variations in a basal sand and how these affect seismic data. The problem is considerably complicated both by more complicated variations in the basal sand and also by changes in nearby formations. Excess pressure in some areas further complicates the problem. On the other hand, the objective is not deep, and the data quality is good. The study concluded that the sand could be seen when thicker than 20 feet (figure 8.13). Inversion of the seismic data to seismic logs (SAILE traces) helped in predicting the distribution and thickness of the sand. Several wells were drilled utilizing these results. In most instances, the predictions were verified, both where the sand was predicted to be too thin and where it was productive. The prediction failed in one well that encountered a different stratigraphic situation than had been considered in the study and that gave nearly the same waveform.

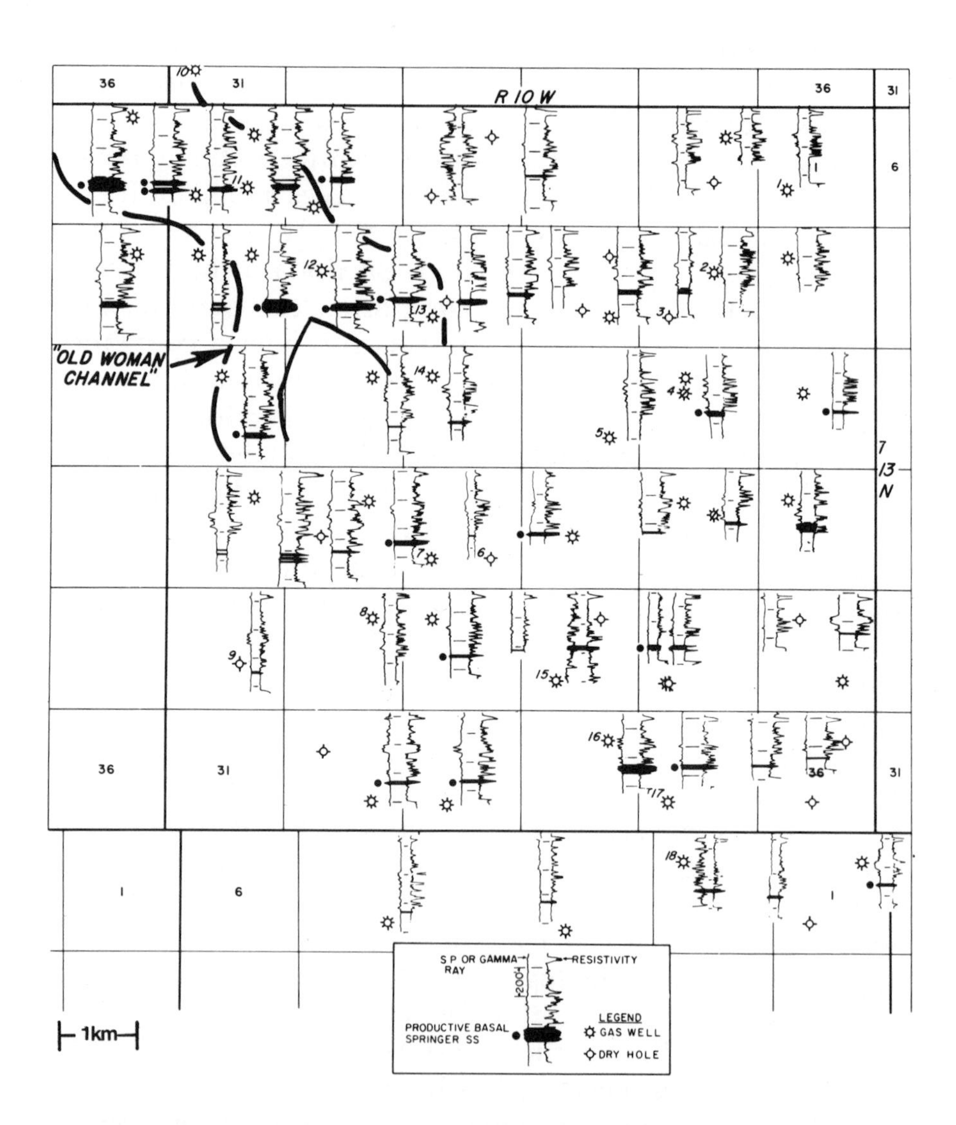

Figure 8.12. Portion of Anadarko Basin selected for Basal Springer sand seismic-stratigraphic analysis, showing locations of wells whose logs were included in study. (From Clement, 1977; reprinted by permission of The American Association of Petroleum Geologists)

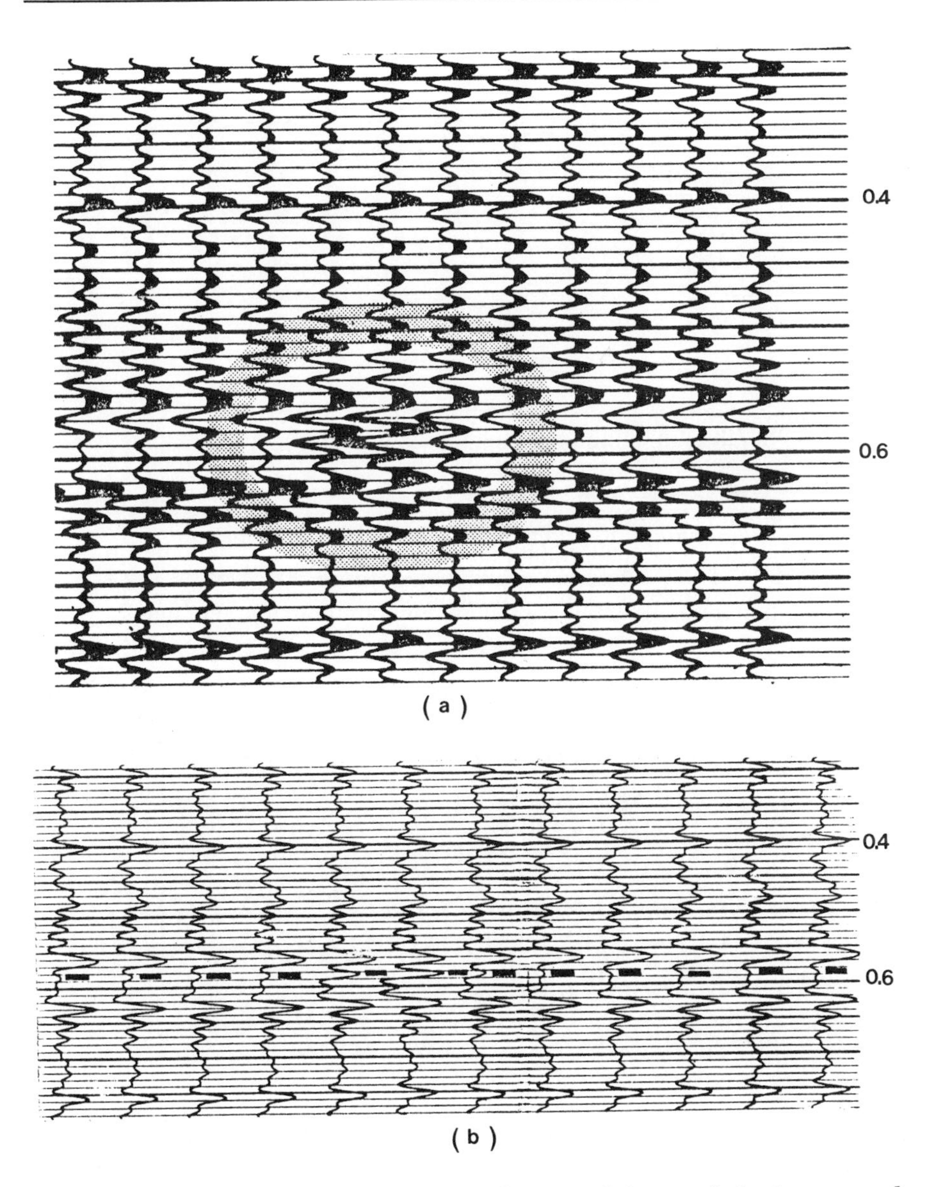

Figure 8.13. Portion of a seismic line used in Basal Springer sand study. (a) Seismic section showing energy build-up at 0.58 s, which indicates greater than 20-foot thickness. (b) Seismic logs (SAILE traces) obtained by inversion of the data shown in (a). (Reprinted by permission of Conoco Incorporated)

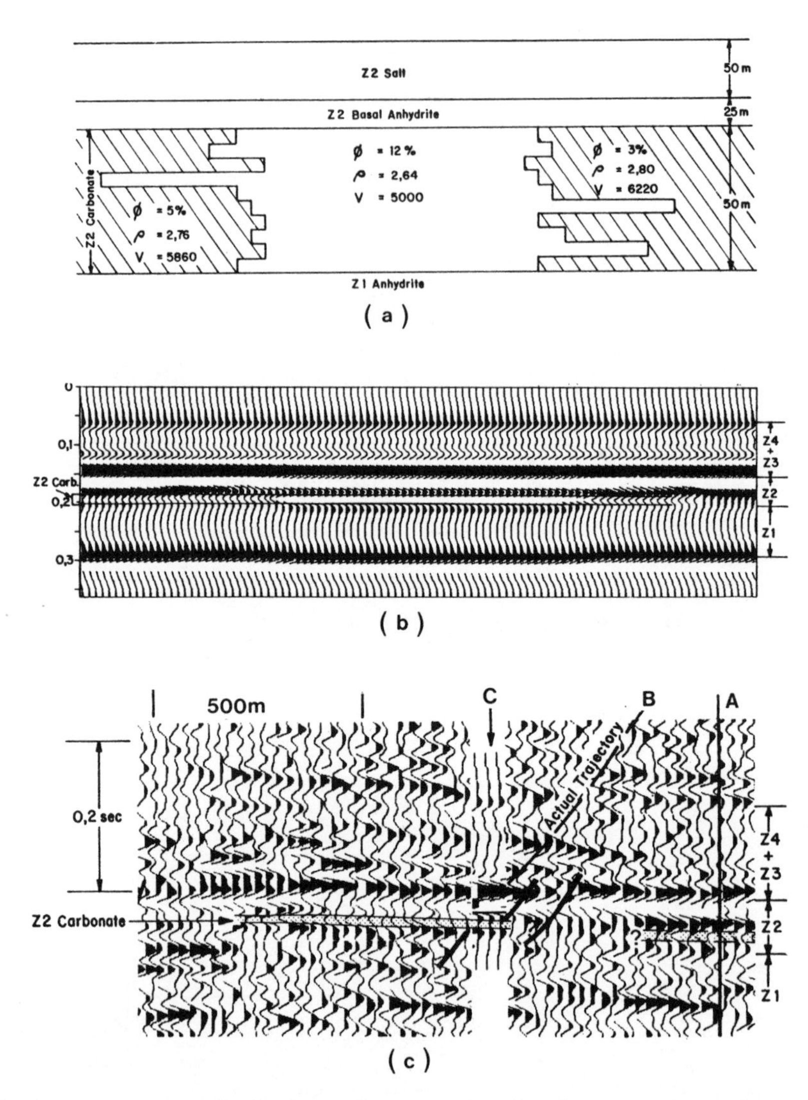

Figure 8.14. Study looking for porosity development in carbonates. (a) Model; (b) synthetic seismogram, with the porous zone being associated with a trough (called "Z2 carb"); and (c) portion of seismic record section through well A that was used to predict what well B (a deviated hole) would find. C is a synthetic seismogram made from well B, which encountered the porous zone. (From Maureau and van Wijhe, 1979; reprinted by permission of The Society of Exploration Geophysicists)

Most seismic modeling studies neglect density as a factor, whether in forward or inverse modeling. Maureau and van Wijhe (1979) presented a case history in which density variation was critical. They endeavored to predict porosity in the Permian Zechstein carbonate of the Netherlands from seismic data. Porous carbonate here has about the same velocity as salt, which is also present, but density differences are present. To get meaningful acoustic impedance logs they used both density (FDC) and sonic (BHC) logs. Figure 8.14a shows a synthetic seismogram that indicates that a trough should develop where porosity is present. Figure 8.14c is a portion of a seismic section showing such a region verified by a well.

Model studies are most valuable when used to extend well control. Hilterman (1977) modeled a stratigraphic trap to see how three-dimensional effects would distort the picture that a grid of seismic lines would give. The solid outline in figure 8.15 shows the trap and the dashed outline as it would be mapped from the seismic lines. In the reentrant areas, the Fresnel zone laps onto the portions of the trap off to the side and so makes the trap appear larger than it actually is. Conventional two-dimensional migration would not remedy such a picture.

Summary of Interpretation Procedures

Reflection character analysis is fundamentally concerned with lateral changes, changes as one follows along a reflection event. The changes are most often in amplitude, waveshape, frequency, velocity, or thickness (interval between events). Synthetic seismograms are made from models of the geology as aids in defining what features of seismic waveform associate with geologic changes. We must realize, however, that agreement between what is predicted from syn-

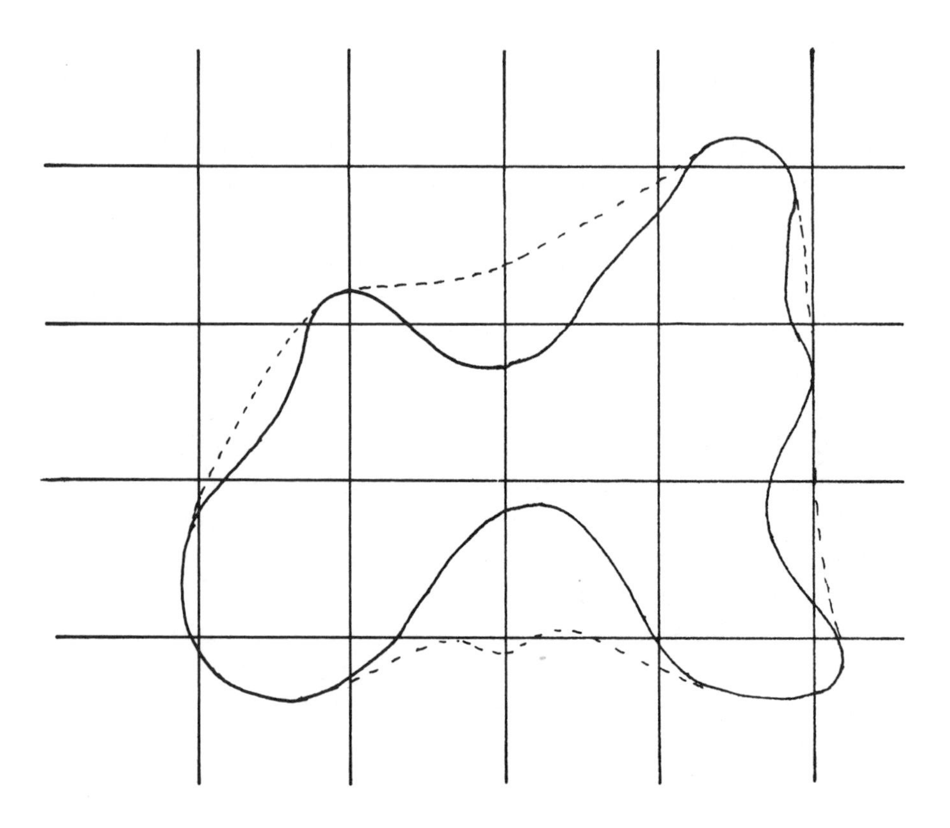

Figure 8.15. Model of a stratigraphic trap (solid line) and grid of seismic lines used to map it. The feature will be seen in reentrant areas because reflections will be seen from the sides where they are within the Fresnel zone. The apparent protrusion on the south reentrant is a focusing effect. (After Hilterman, 1977; reprinted by permission of The American Association of Petroleum Geologists)

thetic seismograms and what is actually seen on seismic records is not a guarantee that the modeled feature agrees with the actual feature, but merely that it is one possibility. Seismic data are also converted to synthetic acoustic-impedance logs in inverse modeling as an aid in seeing the significance of waveshape variations.

9
Hydrocarbon Indicators

Hydrocarbon accumulations sometimes have effects on seismic data that can be used to indicate their location. The recognition of such indications is called *direct detection*, although the line of reasoning is often far from direct. The most prominent of these indicators is often a marked increase in amplitude or a *bright spot* (see figure 1.3), but the entire set of hydrocarbon indicators is sometimes included in the name "bright spot." Bright-spot detection came to the forefront in the early 1970s; it is now regarded as a subset of seismic stratigraphy.

The range of amplitude between early and late reflections on a seismic record exceeds the range of normal displays. This range of amplitude has to be compressed to permit both shallow and deep reflections to be seen, and various data-compression schemes have been used since the early days of seismic exploration. Automatic gain control was used for so long that geophysicists forgot that the magnitude of a reflection carried information, and bright-spot technology was the rediscovery that hydrocarbon accumulations (especially in Tertiary clastic sediments) are often associated with amplitude anomalies.

The presence of hydrocarbons in the pore space of a rock lowers the velocity (see figures 6.7 and 6.8) and also the density. In Tertiary clastic sediments the lowering of the velocity and density is often so large as to produce exceptionally large acoustic-impedance contrasts, resulting in high-amplitude reflections. In other types of sediments, hydrocarbon accumulations may produce sufficient changes that variations in reflectivity along the bedding evidence the accumulation. However, acoustic impedance can change for a number of reasons so that no unique relation exists between amplitude and accumulations.

Seismic Attributes

Measurements of various aspects of the data, called *attributes*, are often displayed in record-section form. Among the attributes useful as hydrocarbon indicators are amplitude, phase, frequency, polarity, and velocity. Complex trace analysis (Taner *et al.*, 1979) provides a method for calculating values for some attributes. Color plate 5 shows attributes displayed as color overlays of a seismic section in the North Sea.

The reflection-strength display (color plate 5a) shows the amplitude of the envelope of the seismic trace. The colors are arranged in a spectral sequence with the brightest (reds and oranges) indicating the largest values and the greens, blues, and whites indicating the smaller ones. The colors are related to numerical values in a quantitative way, each successive color tone indicating 1 dB of amplitude. Where the data consist of separate primary reflections, the amplitude indicates the magnitude of the change in acoustic impedance at the reflecting interface.

The rocks shown in color plate 5 are mainly Mesozoic, but the acoustic-impedance change is sufficient for a gas accumulation in the crest of the anticline to produce a weak amplitude anomaly. Often one of the problems with associating amplitude anomalies with gas accumulations is that small amounts of gas in a rock's pore space produce as large an acoustic impedance change as complete gas saturation (see figure 6.8); hence amplitude bright spots may not indicate enough gas to provide a commercial accumulation.

Phase, as we use it here, is sometimes called *instantaneous phase* to avoid confusion with phase spectra, where phase is a function of frequency. A phase display (color plate 5b) emphasizes the continuity of data. Phase is independent of amplitude so that both weak and strong reflections show up equally. By emphasizing continuity or discontinuity, phase is especially useful in pinning down the locations of angular unconformities (A,B,C) or faults. Phase (or cosine of the phase) is also often displayed in black-and-white form, as in figures 1.10 and 1.11; a phase plot also forms the black background for color plates 5a, 5c, and 5d. The phase display in color plate 5b shows a horizontal reflection (D) presumably resulting from gas-water contact in the reservoir. Such "flat spots" are one of the best hydrocarbon indicators.

Instantaneous frequency (color plate 5c) is sensitive to the interference pattern between closely spaced reflections, and consequently instantaneous frequency often characterizes a particular sequence of reflectors. It is thus especially useful in character-correlating from line-to-line or across faults. It is also sensitive to changes in the bedding sequence and thus is useful in telling where stratigraphic changes occur. The color code for frequency uses the strong colors of red and orange to indicate low frequencies and the colors grade in a spectral sequence to blues indicating the higher frequencies, successive color changes indicating 2 Hz increments.

Hydrocarbon accumulations often have exceptionally low frequencies immediately beneath them; this is referred to as a "low-frequency shadow," and it is often a useful hydrocarbon indicator. The response on color frequency displays is an orange or yellow shade (D in color plate 5c). The distinction between frequency changes because of stratigraphic change and because of hydrocarbon accumulation is provided by the abruptness of the change, the former usually being distributed over a number of traces whereas the latter is usually abrupt. Compare the changes along the unconformities A,B,C with that attributed to the accumulation at D in color plate 5c. Faulting, of course, also produces abrupt change.

An apparent-polarity display (color plate 5d) indicates the sign of the reflection coefficient if a reflection is an isolated primary reflection and if the wavelet is zero phase. The color intensity is modulated by the reflection strength and thus shows the magnitude of the reflection coefficient also. The reflection from the top of the gas reservoir (D) is blue, indicating a negative reflection coefficient, whereas that from the gas-water contact is pink, indicating a positive reflection coefficient.

Other Hydrocarbon Indicators

The low velocity of a hydrocarbon accumulation produces a slight delay in energy that passes through the reservoir and sometimes a velocity sag can be seen in deeper events, although, in order for the effect to be large enough to be seen clearly, the delay has to have been produced by a thick gas accumulation. Other types of velocity effects are occasionally seen also, such as distortion, because of the lensing effect on raypaths that pass through a gas reservoir.

The reflectivity of a bright spot may be sufficiently great to make it an important generator of multiples. A section processed with constant stacking velocity, the stacking velocity for the bright spot (figure 9.1), emphasizes multiples and provides a useful tool for locating the accumulation that generates the family of multiples.

A lowering of amplitudes (amplitude shadow) is often seen underneath accumulations (figure 9.2). This is sometimes attributed to loss of energy in transmission through the accumulation, but the effect more often results from amplitude normalization. Because the reflection from the gas accumulation is so large, the sum amplitude of the trace is higher than that of neighboring traces, and the gain of the trace is decreased to match the sum amplitude of the other traces, making reflections other than those indicating the accumulation appear weaker. The amplitude of events above the accumulation thus may show weakening over an accumulation as well as those below the accumulation. A variety of circumstances other than hydrocarbon accumulations can change reflection amplitude; a few are indicated in figure 9.3. In fact, any of the hydrocarbon indicators can be misleading.

Probably the most reliable indicator is the flat spot— a flat reflection presumably resulting from a gas/oil, gas/water or oil/water interface within a reservoir. However, flat spots are usually not seen distinctly unless the reservoir is more than a 1/4 wavelength thick, so that different reflections associated with the reservoir are at least partially resolved. The flat spot in color plate 5b has just been noted, and a flat spot can also be seen in figure 1.11.

The lowering of velocity in a hydrocarbon accumulation sometimes changes the reflection coefficient at the top of a reservoir from positive to negative, resulting in polarity reversal (figure 9.4). More commonly, the reflection from the accumulation is only one component of a composite reflec-

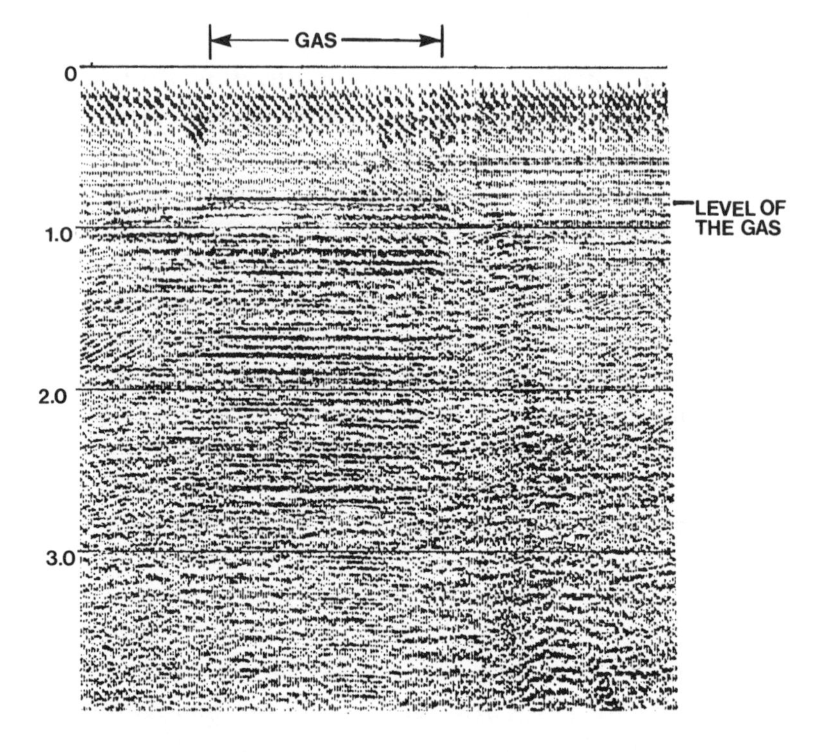

Figure 9.1. Multiples generated by a gas sand. The section is stacked at the stacking velocity appropriate to the gas sand at about 0.8 s. (From Sheriff, 1975; reprinted by permission of The European Association of Exploration Geophysicists)

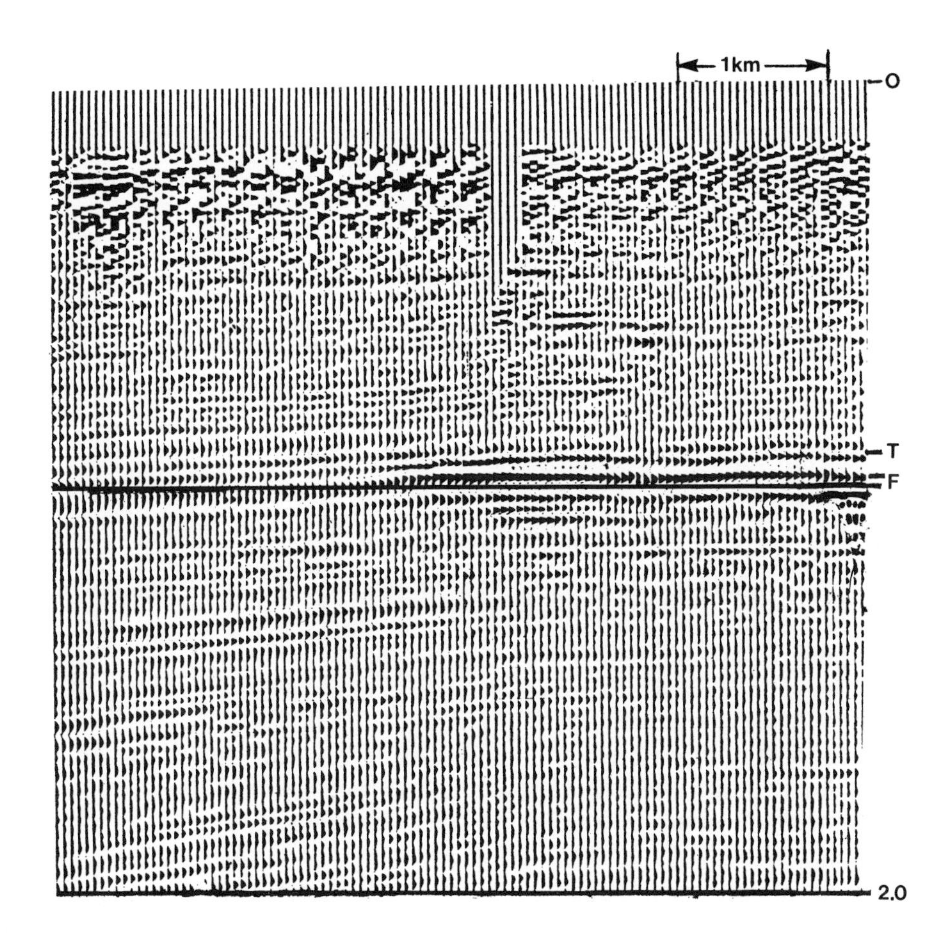

Figure 9.2. Seismic section showing a gas sand. The top of the reservoir is at T; F is a flat spot from the gas/water contact. The events associated with T and F are 180° out of phase. Note the weakened reflections underneath the gas sand. (From Sheriff, 1975; reprinted by permission of The European Association of Exploration Geophysicists)

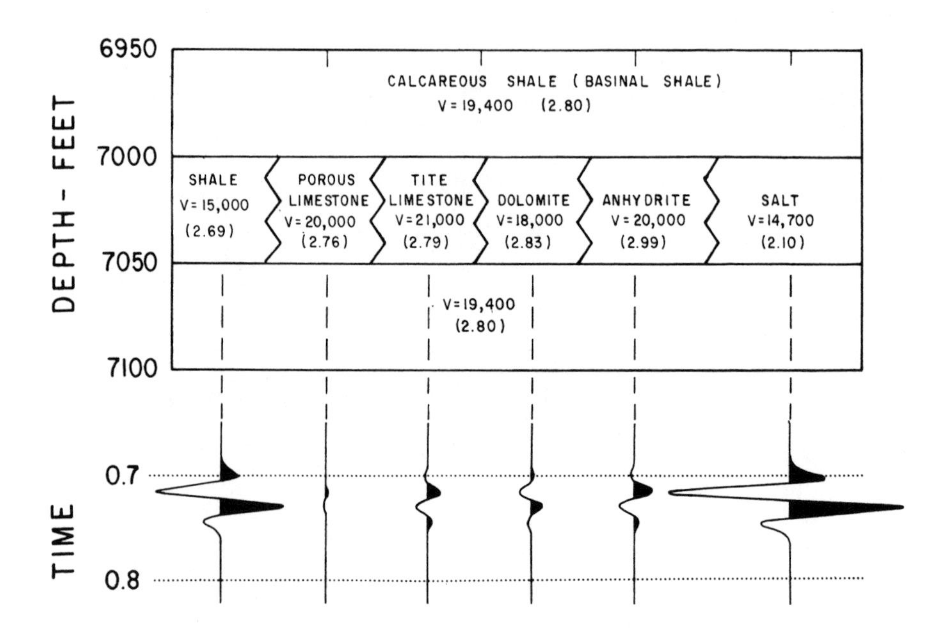

Figure 9.3. Reflectivity of various rock layers imbedded in a calcareous shale. (From Meckel and Nath, 1977; reprinted by permission of The American Association of Petroleum Geologists)

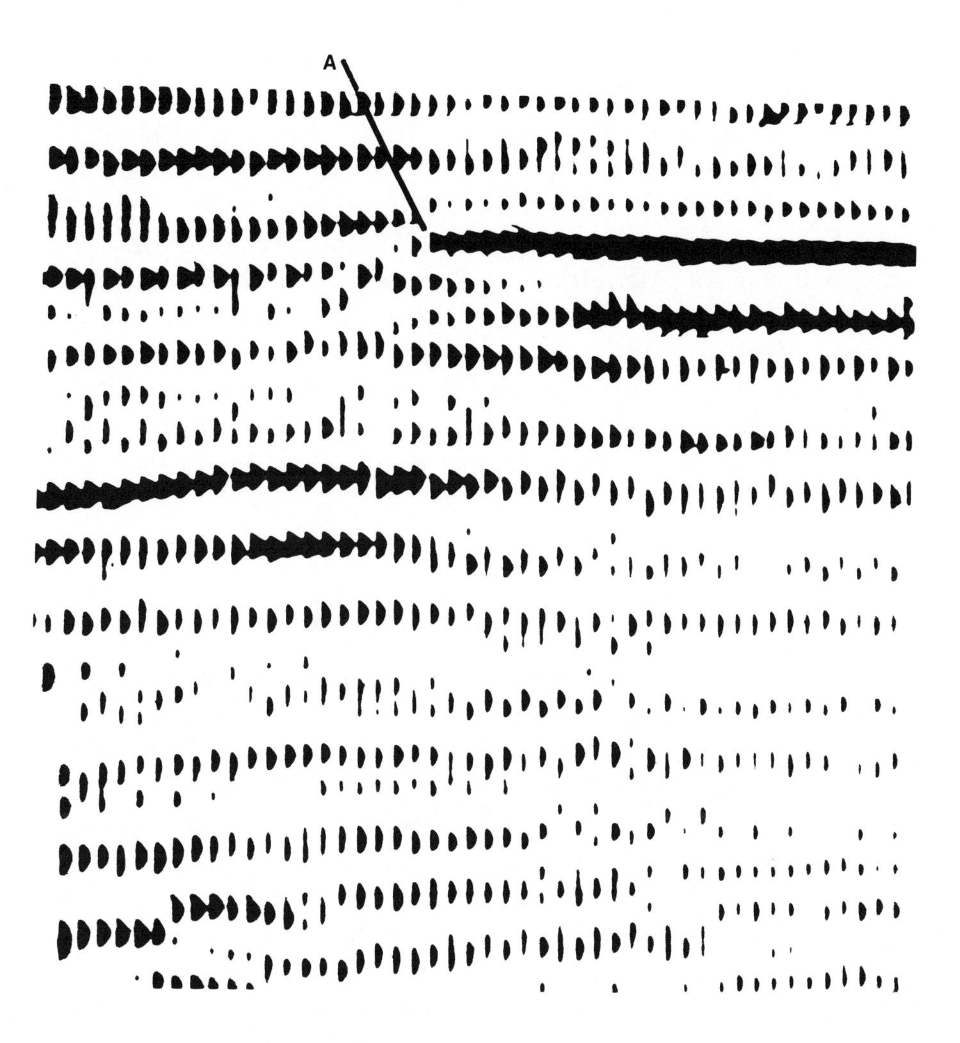

Figure 9.4. Polarity change at edge of a gas accumulation (A). Note decrease in amplitude of reflections both underneath the reservoir and also over it.

tion so that a change in this component changes the phase of the composite. Sometimes the lowering of velocity results in the contrast against capping sediments becoming smaller, so that accumulation is evidenced by a weakening of the amplitude, producing a *dim spot* (figure 9.5).

No hydrocarbon indicator by itself is foolproof. A conjunction of indicators provides a much more reliable indication of an accumulation than any individual indicator. Thus, strong amplitude with a low-frequency shadow at a location where trapping conditions might exist (in the crest of an anticline or with a fault at the up-dip edge, for example), can together provide a strong case. If there should be a flat spot in addition, the case is very strong.

Examples of Color-Attribute Interpretation

Color plate 3 is an enlargement of a frequency display in the Williston Basin where accumulations occur in the meanders of ancient stream channels. The ancient surface is marked UU′. The channel indicated is productive. Channels show up much more clearly on frequency color displays than on conventional displays, but they can also be seen on conventional displays. Enlarging the portion of the section where the channels occur permits one to see detail otherwise not obvious. Anomalies are seen on this section in other places also, but there is sufficient well control available to allow one to concentrate attention on the specific reflection zone.

An amplitude section across a structure offshore Africa is shown in color plate 4. Predictions of hydrocarbons were made based on both amplitude and frequency displays, and a well was drilled on the crest of the structure. Hydrocarbon-bearing zones in the well are indicated by the red regions on the vertical bar that indicates the well. The bright spot A was noncommercial gas. Most of the remaining zones are prob-

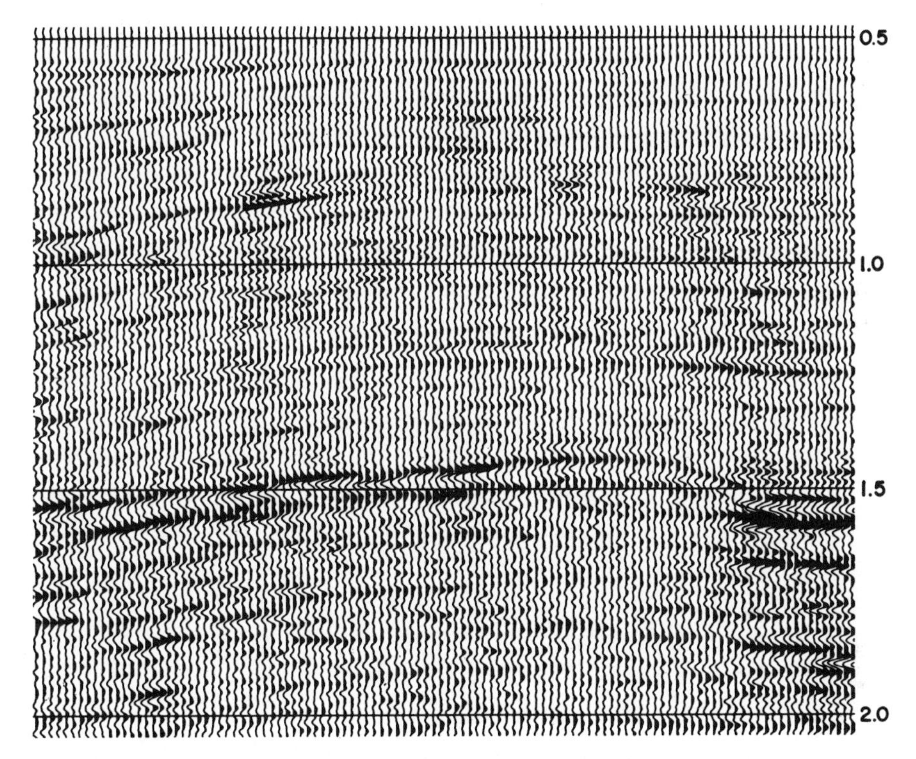

Figure 9.5. "Dim spot" resulting from gas accumulation in porous carbonates with shale cap. (Reprinted by permission of Teledyne Exploration)

ably commercial and most correlate with either amplitude or frequency anomalies, or with both. The deepest accumulation (F) was not predicted from this study.

Color plate 6 is of a line in the Gulf of Mexico. A number of amplitude anomalies seem to be associated with what is apparently a fault. These involve gas but not all are commercial. The association with the fault suggests that the gas migrated up the fault plane to reach these reservoirs. A major accumulation of condensate lies at D. It produces only a weak amplitude anomaly but a substantial frequency anomaly. The shallow amplitude (and frequency) anomalies might be thought of as "pointers" to the deeper accumulation. The overall uplift is probably due to deep salt flowage.

Defining Reservoir Limits

Seismic data can be especially valuable in extending well information to the surrounding region. Many uncertainties are reduced significantly by well control. One application of hydrocarbon-indicator analysis is in delineating the limits of reservoirs for use in field development. Attributes can show faults and other features that separate different pools and aid in mapping the areal extent of various pools.

Color plate 7 shows a line in the Gulf of Mexico where a number of gas reservoirs were found. The gas reservoirs have associated amplitude and low-frequency anomalies where the reservoirs exceed about 10 feet in thickness. Faults that can be seen clearly in the seismic data separate the field into a number of separate pools. The limits of the pools as determined by the seismic data and by subsequent drilling is in good agreement. Reservoir M is 12,000 feet deep.

Color plate 8 shows a seismic line offshore Europe where production comes from carbonate rocks. This line shows two anticlines. The right anticline produces both gas and oil while the left is barren in the corresponding intervals. The limits of gas and oil production were predicted based on the seismic data. Subsequent drilling supported the predicted gas limits, but the oil predictions turned out to be incorrect. The oil distribution in this field is erratic; oil producibility limits are apparently controlled in different places by porosity development, gradual water-saturation variations, or an oil/water contact.

The barren structure to the left clearly appears different in attribute appearances. While the seismic data did not everywhere predict the oil limits correctly, they did assist in the field development.

Summary of Hydrocarbon Indicators

While no single attribute provides a reliable indicator of hydrocarbons by itself, a conjunction of some of the following indicators provides strong evidence:

(1) Amplitude change, a brightening or dimming of the reflection amplitude; amplitude also affects the generation of multiples because of the change in the reflection coefficient.

(2) Frequency change, generally involving a lowering of frequency immediately below the reservoir.

(3) Velocity change, almost always involving a lowering of velocity; where the reservoir is sufficiently thick, it may produce a velocity anomaly in reflections beneath it.

(4) A change in wave shape, which sometimes shows up as a reversal of polarity and sometimes by other phasing changes.

(5) A flat spot that can result when a reservoir is sufficiently thick.

(6) Various other things, such as association with a trapping mechanism (reservoir being located on an anticline crest or at a fault), a presence of associated accumulations, indications of gas leaking out of a reservoir, and so on.

Envoi

We have tried to show some of the techniques that have been developed for determining the stratigraphy from seismic data in areas (1) where we lack a stratigraphic framework and (2) where we know the stratigraphy in a general way but wish to locate variations of exploration interest. We wish to encourage applications of seismic stratigraphy to develop the art further.

> We must believe . . . but at the same time we must be thoroughly convinced that we know this relation only in a more or less approximate way, and that the theories we hold are far from embodying changeless truths. When we propound a general theory in our sciences, we are sure only that, literally speaking, all such theories are false. They are only partial and provisional truths which are necessary to us, as steps on which we rest, so as to go on with investigation; they embody only the present state of our knowledge, and consequently they must change with the growth of science, and all the more often when sciences are less advanced in their evolution. (Claude Bernard in *Experimental Medicine.*)

Certainly seismic stratigraphy qualifies as a "less advanced" science. Let us get on with investigation.

Glossary

acoustic impedance: the product of seismic velocity and density.

aggradation: building upward by deposition (see *onlap*).

anti-alias filter: a filter used before sampling to remove high frequencies that otherwise would cause ambiguities in information content. "Alias" and "anti-alias" filter mean the same thing.

apparent polarity: see *polarity*.

apparent velocity: the velocity that a waveform appears to have in some direction, usually in the horizontal direction and usually not the direction of the wave.

attributes (seismic): measurements of properties of seismic data, such as amplitude, frequency, velocity, polarity, and so on.

back-reef facies: deposits on the landward side of a reef.

baselap: angular termination of strata at the lower boundary of a depositional sequence; includes both *onlap* and *downlap*.

biogenic rock: rock manufactured directly by organisms.

bioherm: a mound-like or reef-like mass of rock built up by sedentary organisms.

bright spot: (1) a local build-up of amplitude, which sometimes indicates a hydrocarbon accumulation; (2) the class of hydrocarbon indicators, one of which is an amplitude anomaly.

carbonate-build-up configuration: a seismic-reflection configuration, usually a mound, interpreted as a carbonate reef or bank.

chaotic configuration: discontinuous, discordant seismic-reflection pattern interpreted either as strata deposited in a highly variable, high-energy setting, or as strata that have been so deformed that continuity has been disrupted.

chronographic/chronostratigraphic chart: a summary chart on which geologic time is plotted vertically and distance horizontally and on which stratigraphic information is brought together (see figure 3.12).

clastics: sediments composed of fragments transported from their point of origin.

clinoform surface: a sloping depositional surface, commonly associated with strata prograding into deep water.

coastal aggradation: see *coastal onlap.*

coastal encroachment: see *coastal onlap.*

coastal deposits: sediments deposited near sealevel (littoral, paralic, or coastal nonmarine).

coastal onlap: the progressive landward onlap of coastal (littoral or coastal nonmarine) deposits. Coastal aggradation and coastal encroachment are the vertical and horizontal components of coastal onlap, respectively (see also *onlap*).

coherent: having a fixed phase relationship to each other.

coherence filter: a multichannel filter that enhances coherent events.

common-depth point stacking: the combination of data that have in common the same midpoint between source and detector.

competent rock: a rock that maintains its shape compared to adjacent rocks when subjected to a tectonic force.

compressional wave: a seismic wave in which particle motion is in the direction of the wave.

concordance: parallelism of strata to sequence boundaries (see figure 3.11).

constructive shelf: a shelf where sediment influx exceeds marine erosional effects (see figure 5.6).

convergent reflections: see *divergent reflections.*

convolution: a mathematical process constituting linear filtering. Natural processes in the earth appear to accomplish convolution (see *convolutional model*) and many instrumental and processing procedures involve convolution. "Linear superposition" constitutes convolution (see also *deconvolution*).

convolutional model: the concept that a seismic trace can be represented by the convolution of an equivalent wavelet and the earth's reflectivity, with noise added.

cycle of relative change of sealevel: the interval of time during which a relative rise and fall of sealevel takes place. Three orders of cycles are recognized: first-order, supercycle, and cycle (see also *paracycle, eustatic cycle,* and figure 3.4).

deconvolution: removing the effect of an earlier filtering action. "Predictive" deconvolution involves the prediction and removal of multiples that involve the near-surface as one of the reflectors. "Spiking" deconvolution attempts to achieve an equal amplitude spectrum at all frequencies and sharpen the effective wavelet. *Wavelet processing* can be thought of as a type of deconvolution.

depositional sequence: see *time-stratigraphic unit.*

destructive shelf: a shelf where the effects of marine erosion exceed those of sediment influx (see figure 5.6).

detectable limit: the minimum thickness for a bed to give a seismic reflection, in the vicinity of 1/30 of the dominant wavelength.

differential compaction: uneven settling of sediments as a result of reduction of porosity. Sands/shales generally shrink more under pressure than reef limestones (see figure 5.16).

diffraction: the energy returned from a point, such as the abrupt termination of a reflector at a fault, unconformity, reef, salt dome, and so on. Diffraction energy does not satisfy Snell's law at the diffracting point.

dim spot: a local lowering of amplitude, sometimes because of hydrocarbons being locally present in a rock where the acoustic impedance is appreciably greater than that of the overlying rock (see figure 9.5).

direct hydrocarbon indicator: something in seismic data that may indicate an accumulation of gas or oil.

direct modeling: see *forward modeling.*

distal onlap: see *onlap.*

divergence correction: correction applied to amplitude values to compensate for the decrease in energy density with distance from the source, that is, to compensate for geometrical spreading.

divergent reflections: seismic reflections characterized by a laterally thickening wedge-shaped unit in which thickening is accomplished by splitting of individual reflection cycles within the unit, rather than thickening predominantly by onlap at the base or toplap or erosion at the top of the unit.

downlap: a relation in which initially inclined strata terminate downdip.

downward shift of coastal onlap: a seaward shift from the highest position of coastal onlap in a seismic sequence to the first onlap in the next younger unit. It is used to recognize relative falls of sealevel.

drape: a unit draped evenly over a preexisting surface without regard for the preexisting topography.

earth reflectivity: the time series that is the sum of the reflection coefficients at all interfaces spaced according to reflection times. Also called a "stickogram" (see figure 7.1).

encroachment: see *coastal onlap.*

erosional truncation: see *truncation.*

eustatic cycle: the interval of time during which a worldwide rise and fall of sealevel takes place (see figure 3.4).

eustatic level chart: a chart showing worldwide changes in sealevel (see figure 3.4).

facies: the sum of all characteristics exhibited by a sedimentary rock and from which its origin and environment of formation may be inferred; the general aspect, nature, or appearance of a sedimentary rock produced under or affected by similar conditions; a distinctive group of characteristics that differs from other groups within a stratigraphic unit.

facies (seismic): see *seismic facies.*

fill configuration: seismic reflections interpreted as strata filling a negative relief feature. Fill units are classified by external form (channel fill, trough fill, basin fill, slope-front fill; (see figures 5.7 and 5.8) or by internal reflection configurations (onlap, mounded onlap, divergent, prograding, chaotic, complex).

first-order cycle: a cycle of relative or eustatic change in sealevel that has a duration in the order of 200 to 400 million years (see figure 3.4).

fluvial: relating to a river, as with deposits laid down by a stream.

fondoform: the seaward portion of a progradational pattern; foreset.

fore-reef facies: sediments laid down seaward of a reef.

forward modeling: computing the effects of a model, as in *synthetic seimogram* manufacture.

frequency spectrum: the amplitude and phase of sine waves of different frequencies that would add up to a particular waveform.

Fresnel zone: the portion of a reflector responsible for a reflection event seen at a point. Reflection ray paths from source to detector that differ by less than ½ wavelength interfere constructively; the portion of the reflector from which they can be reflected constitutes a "Fresnel zone."

Gardner's relation: the relationship that density is proportional to the ¼ power of seismic velocity (see figure 6.2).

hemipelagic facies: deep-sea sediments, mostly very fine-grained.

hiatus: an interval of geologic time not represented by strata at a specified position. If the hiatus encompasses a measurable interval of geologic time, the surface is an "unconformity." A "nondepositional" hiatus refers to an interval during which no strata were deposited, an "erosional" hiatus to an interval during which strata were removed by erosion.

highstand: a time when sealevel is above the shelf edge.

hinterland sequence: a sequence of nonmarine deposits laid down at a site interior to the coastal area, where depositional mechanisms were not controlled by sealevel.

hummocky clinoform: a reflection configuration that consists of irregular, discontinuous, subparallel reflections forming a practically random, hummocky pattern. This pattern commonly grades laterally into larger, better-defined clinoform patterns, and upward into parallel reflections.

Hz: hertz or cycles per second.

imaging: see *migration.*

impulsive source: a source of very short duration.

interference: the superposition of waveforms is "constructive" where they add in-phase and "destructive" where they add out-of-phase.

inverse modeling: calculating a model from observed effects (see *seismic log*).

inversion: conversion of reflectivity to acoustic-impedance values (see *seismic log*).

instantaneous frequency: the time rate of change of phase. Instantaneous frequency is a tool for seeing lateral stratigraphic or hydrocarbon variations.

instantaneous phase: see *phase.*

instantaneous velocity: see *seismic log.*

isochronous surface: see *time surface.*

isopach: a line drawn through points of equal thickness for a unit.

isotime: (1) sometimes used for *isopach* where thickness is measured in reflection-time difference. (2) Isotime surface is sometimes used for a surface of constant time (see *time surface*).

linear inverse model: a type of migrated section (see figure 2.7).

littoral: relating to the depth zone between high and low water.

lowstand: a time when sealevel is below the shelf edge.

marine onlap: see *onlap.*

migrating wave: a reflection configuration consisting of superimposed wave-shaped reflections, each progressively offset laterally from the preceding.

migration: repositioning of reflected energy so that it indicates the location of the reflecting point; also called "imaging."

minimum phase: a characteristic of waveforms that have their energy concentrated early in the waveform (see figure 6.10).

model: a concept of a relationship between cause and effects.

mound configuration: a reflection configuration interpreted as strata forming an elevation above the general level of surrounding strata (see figure 5.7).

multiples: energy that is reflected more than once.

neritic: relating to water depths between low tide and 200 m.

noise: any unwanted energy.

nondepositional hiatus: see *hiatus.*

normal incidence: the situation where the direction of travel of a wave is perpendicular to an interface.

normal moveout: differences in the arrival time of reflections because of the distance between source and receiver.

oblique reflection: a reflection configuration interpreted as a prograding (clinoform) pattern whose upper surface is near the wave base. Strata terminate updip by toplap and downdip by downlap. Successively younger foreset segments build laterally from a relatively constant upper surface characterized by lack of top-set strata (see figures 5.4 and 5.5). The upper part of oblique sequences are apt to be sand prone.

offlap: a pattern in which each successively younger unit leaves exposed a portion of the older unit on which it lies, interpreted as strata prograding into deep water.

offset: the distance between source and detector.

onlap: a base-discordant relationship in which initially horizontal or inclined strata terminate progressively updip against a surface of greater initial inclination. "Proximal" onlap is onlap in the direction of the source of clastic supply; "distal" onlap is in the direction away from the source; "marine" onlap is onlap of marine strata (see figure 5.6); "coastal" onlap is the progressive landward onlap of coastal deposits.

onlap fill: see *fill configuration.*

outbuilding: see *progradation.*

out-of-phase: a relation between two waveforms where the phase differs, usually by nearly 180°.

paracycle: the interval of time occupied by a rise and stillstand of sealevel, followed by another rise, with no intervening fall.

parallel configuration: parallel seismic reflections within a seismic sequence.

pelagic deposits: sediments containing little of continental source, being mainly from material in general oceanic suspension.

phase: (1) the amount of rotation in circular motion, or the stage in the cycle of a harmonic or pseudoharmonic waveform. Phase carries the timing information in seismic data. (2) An "instantaneous phase" plot shows phase (often in color-coded form) or some function of the phase (such as the cosine of the phase) as a function of arrival time. A phase plot emphasizes the continuity of coherent events or abrupt interruptions of continuity.

polarity: the sign of the reflectivity (positive or negative) that indicates whether the acoustic impedance is increasing or decreasing at an interface. Apparent polarity is the polarity if a single interface is involved and if the effective wavelet is zero phase.

predictive deconvolution: a process that uses early arrivals to predict and remove later multiples involving the same reflectors.

prodelta: the seaward part in front of a delta complex, lying below the wavebase.

progradational configuration: seismic reflections interpreted as deposition due to lateral outbuilding (and sometimes upbuilding). Progradational reflection patterns include sigmoid, oblique, complex, and so on.

proximal onlap: see *onlap.*

reef: a ridge-like or mound-like structure built by sedentary organisms.

reflection character analysis: determining stratigraphy or hydrocarbon accumulations from the detailed seismic waveshape (see chapter 8).

reflection coefficient: see *reflectivity.*

reflection configuration: the geometric pattern of seismic reflections, interpreted to represent configuration of strata.

reflection-free: absence of reflections, interpreted as due to homogeneous, nonstratified, highly contorted, or very steeply dipping geologic units.

reflectivity: the ratio of the amplitude of a reflected wave to that of the incident wave. For normal incidence (direction of wave perpendicular to interface), the reflectivity is given by the ratio of the change in acoustic impedance to twice the average acoustic impedance. Also called "reflection coefficient." Reflectivity is generalized to include the effects of a series of interfaces (see *earth reflectivity*).

regression: a seaward movement of the shoreline indicated by the littoral facies moving seaward with time (see figure 3.2b).

relative change in sealevel: an apparent rise or fall of sealevel with respect to the land surface. Either sealevel rises or the land surface subsides, or both.

relative fall of sealevel: an apparent fall of sealevel with respect to the surface of deposition. Recognized by a downward shift of coastal onlap (see figure 3.3).

relative rise of sealevel: an apparent rise of sealevel resulting from sealevel rising or the land subsiding. Recognized by coastal onlap (see figure 3.2)

relative stillstand of sealevel: an apparent constant stand of sealevel with respect to the surface of deposition. Recognized by coastal toplap and lack of coastal onlap.

resolvable limit: the minimum thickness of a bed for the effects of the bed's upper and lower surfaces to give distinguishable reflection effects, considered as ¼ of the dominant wavelength.

resolution: the limit at which the effects of two features can be distinguished from the effects of one feature.

reverberation: energy that bounces back and forth within a layer, especially a water layer.

Ricker wavelet: a particular zero-phase wavelet of convenient mathematical formulation, which is often used in synthetic seismogram studies.

ringing character: a wavelet that has more cycles than usual.

SAILE trace: see *seismic log.*

sand-prone seismic facies: seismic facies interpreted as deposited in a clastic environment with sufficiently high energy to sort sand (assuming sufficient quantities of sand-sized particles were available).

second-order cycle: see *supercycle.*

seismic facies: distinctive features that characterize seismic reflections, such as continuity, amplitude, configuration, and so on.

seismic-facies analysis: the description and geologic interpretation of seismic-reflection parameters, including configuration, continuity, amplitude, frequency, and interval velocity, and interpretation as to the implied depositional environment (see chapter 5).

seismic-facies map: a map that shows the areal distribution, configuration, thickness or other aspect of a given seismic-facies unit or parameter.

seismic-facies unit: a mappable three-dimensional unit of reflections whose characteristics differ from those of adjacent facies. Sometimes used interchangeably with "seismic sequence."

seismic log: acoustic impedance or velocity calculated from observed seismic data by inversion, making certain assumptions. Also called "SAILE trace," "synthetic acoustic-impedance log," "synthetic sonic log," or "instantaneous velocity."

seismic sequence: a depositional sequence identified on a seismic section by mapping the bounding unconformities. Sometimes used interchangeably with "seismic facies unit."

seismic-sequence analysis: the identification of depositional sequences by subdividing a seismic section into packages of concordant reflections separated by surfaces of discontinuity, and interpreting them as depositional sequences (see *sequence*).

seismic stratigraphy: the study of stratigraphy and depositional facies as interpreted from seismic data.

sequence: a term applied to a relatively conformable succession of genetically related strata bounded at top and base by unconformities or correlative conformities. A "depositional" sequence *(time-stratigraphic unit)* is a stratigraphic unit. A "seismic" sequence is a depositional sequence identified on a seismic section.

shale-prone seismic facies: seismic facies interpreted as clastics deposited where depositional energies are insufficient to produce significant sorting.

shear wave: a seismic body wave in which particles move at right angles to the direction of the wave.

sheet: a thin, widespread, tabular unit.

shingled configuration: a prograding seismic pattern with parallel upper and lower boundaries and very gently dipping, parallel, oblique, internal reflections that terminate by apparent toplap and downlap.

sigmoid configuration: a prograding clinoform pattern formed by superposed sigmoid (S-shaped) reflections, interpreted as strata deposited in deep, quiet water (see figures 5.4 and 5.5). Usually shale prone.

signature correction: a process to change observed seismic data into what would have been obtained with a desired wavelet shape, used where the wavelet shape was known.

simulated sonic log: see *seismic log.*

singing record: a record that involves reverberation effects. See *reverberation.*

sonic log: a record of the transit time over an interval (the reciprocal of velocity) measured in a borehole.

stacking: combining the data from different records.

stacking velocity: the velocity that produces the optimum stack for certain events. Stacking velocity is determined in velocity analysis. For horizontal velocity layering, the stacking velocity is a root-mean-square average of the velocities. Stacking velocity is often called erroneously "rms velocity" where the foregoing assumption is not satisfied.

static correction: a correction for variations in arrival time because of near-surface thickness, velocity, elevation, or datum variations.

stickogram: see *earth reflectivity* and figure 7.1.

stillstand: see *relative stillstand of sealevel.*

stratal configuration: the geometric patterns and relations of strata within a stratigraphic unit, indicative of depositional environment (and later structural movement).

stratal surfaces: surfaces that separate sedimentary strata (see *time surface*). They may represent short periods of nondeposition or a change in the depositional regime.

stratigraphy: "the branch of geology that deals with the definition and description of major and minor natural divisions of rocks . . . and with the interpretation of their significance in geologic history, specifically the geologic study of the form, arrangement, geographic distribution, chronologic succession, classification, and especially correlation and mutual relationships of rock strata . . . in terms of their origin, occurrence, environment, thickness, lithology, composition, fossil content, age, history, paleogeographic condition, relation to organic evolution, and relation to other geologic concepts." (AGI, *Glossary*, 1972)

stratum: a tabular layer of sedimentary material visually separable from adjacent layers by a change in character. A general term that includes both "bed" and "lamination." The term is more frequently used in its plural form, "strata."

supercycle: a group of cycles of relative change in sealevel in which a cumulative rise to a higher position of sealevel is followed by a cumulative fall to a lower position. The duration of a supercycle is 10 to 80 million years. A "second order cycle."

superposition: the principle that the result of an operation is the same as the sum of the results of the operation carried out in pieces (see *convolution*).

surface-consistent: assignment of time delays or amplitude attenuation to source or detector locations.

synthetic seismogram: a seismic trace or section calculated from a model.

synthetic sonic log: see *seismic log.*

third-order cycle: a cycle of change of sealevel that has a duration of 1 to 10 million years, the fundamental cycle that represents only one rise and fall; a "cycle."

time-average equation: an empirical relation for the velocity in a two-component system, the two components usually being the matrix and the fluid in the matrix's pore space. If V_m is the velocity of the matrix material, V_f the velocity of the fluid, and ϕ the porosity, then the velocity $V = (1 - \phi)/V_m + \phi/V_f$.

time-stratigraphic unit: a three-dimensional set of facies deposited contemporaneously as parts of the same system, genetically linked by depositional processes and environments. Also called a "depositional sequence."

time surface: a surface that at one time was the surface of the solid earth. Strata and seismic reflections have the attitude of time surfaces. Also called isochronous and stratal surface.

toplap: termination of strata against an overlying surface mainly as a result of nondeposition (sedimentary bypassing) with only minor erosion.

two-dimensional filtering: the attenuation of wavetrains based on their apparent velocity or apparent dip; a type of coherence filtering.

transgression: a landward movement of the shoreline indicated by landward migration of littoral facies (see figure 3.2a).

transit time: the reciprocal of seismic velocity.

truncation: termination of strata or seismic reflections along an unconformity surface due to postdepositional erosion.

turbidite: a sediment deposited by a turbidity current.

turbidity current: a density current of water with other matter in suspension that sometimes flows swiftly down submarine slopes.

unconformity: a surface of erosion or nondeposition that separates younger strata from older rocks and represents a signifiant hiatus (at least a correlatable part of a unit) not represented by strata. Unconformities occur at each global fall of sealevel, although in some areas of continuous deposition, the hiatus may be too small to detect seismically.

undaform: the subaqueous landform produced by wave action.

upbuilding: aggradation (see *progradation*).

velocity: the speed with which seismic waves travel.

velocity analysis: a procedure for determining the stacking velocity (or normal moveout) that will optimize certain events on a stacked section.

velocity anomaly: a distortion of seismic data (especially of arrival times) because of changes in velocity in the horizontal direction (see figure 5.16).

velocity filtering: attenuating events on the basis of dip moveout.

wave-equation migration: a processing method to migrate seismic data by downward continuation of the wavefield by numeric solution of the wave equation (see *migration*).

wavelet compression: processing with the objective of producing the seismic section that would result if the wavelet had been very short (ideally, impulsive).

wavelet processing: processing that involves determining the seismic wavelet shape and/or changing the effective wavelet shape. Wavelet processing can mean any of a number of different processes with different objectives.

Wiener filtering: processing so as to change a waveform into some desired waveform as nearly as possible in a least-squares sense.

zero phase: a characteristic of waveforms that are symmetrical (see figure 6.10).

References

A.G.I., *Glossary of Geology* (1972), American Geological Institute, Falls Church, Virginia.

Bates, R.L., and Jackson, J.A. (1980), *Glossary of Geology* (2nd ed.), American Geological Institute, Falls Church, Virginia.

Becquey, M., Lavergne, M., and Willm, C. (1979), Acoustic impedance logs computed from seismic traces: *Geophysics* v. 44, p. 1485-1501.

Boise, S. (1978), Calculation of velocity from seismic reflection amplitude: *Geophysical Prospecting* v. 26, p. 163-174.

Brown, L.F., and Fisher, W.L. (1979), Principles of seismic stratigraphic interpretation: Notes for AAPG Petroleum Exploration School.

Brown, L.F., and Fisher, W.L. (1977), Seismic-stratigraphic interpretation of depositional systems: examples from Brazilian rift and pull-apart basins: *AAPG Memoir 26,* p. 213-248.

Bubb, J.N., and Hatlelid, W.G. (1977), Seismic recognition of carbonate buildups: *AAPG Memoir 26,* p. 185-204; also in *AAPG Bulletin* v. 62, p. 772-791.

Clement, W.A. (1977), Case history of geoseismic modeling of basal Morrow-Springer Sandstones, Watonga-Chickasha trend, Oklahoma: *AAPG Memoir 26,* p. 451-476.

Dobrin, M.B. (1977), Seismic exploration for stratigraphic traps: *AAPG Memoir 26,* p. 329-352.

Domenico, S.N. (1974), Effect of water saturation on seismic reflectivity of sand reservoirs encased in shale: *Geophysics* v. 39, p. 759-769.

Galloway, W.R., Yancey, M.S., and Whipple, A.P. (1977), Seismic stratigraphic model of depositional platform margin, Eastern Anadarko Basin, Oklahoma: *AAPG Memoir 26,* p. 439-450; also in *AAPG Bulletin* v. 61, p. 1437-1447.

Gardner, G.H.F., Gardner, L.W., and Gregory, A.R. (1974), Formation velocity and density—the diagnostic basics of stratigraphic traps: *Geophysics* v. 39, p. 770-780.

Garotta, R., and Michon, D. (1967), Continuous analysis of the velocity function: *Geophysical Prospecting* v. 15, p. 584-597.

Gregory, A.R. (1977), Aspects of rock physics from laboratory and log data that are important to seismic interpretation: *AAPG Memoir 26,* p. 15-46.

Hilterman, F.J. (1977), Seismic velocities, bright spot and modeling: Notes for AAPG Petroleum Exploration School.

Hun, F. (1978), Correlation between seismic reflection amplitude and well productivity—a case study: *Geophysical Prospecting* v. 26, p. 157-162.

Lavergne, M., and Willm, C. (1977), Inversion of seismograms and pseudo-velocity logs: *Geophysical Prospecting* v. 25, p. 231-250.

Kerr, R.A. (1980), Changing global sealevels as a geologic index: *Science* v. 209, p. 483-486.

Lindseth, R.O. (1979), Synthetic sonic logs—a process for stratigraphic interpretation: *Geophysics* v. 44, p. 3-26.

Lyons, P.L., and Dobrin, M.B. (1972), Seismic exploration for stratigraphic traps: *AAPG Memoir 16,* p. 225-243.

Marr, J.D. (1971), Seismic stratigraphic exploration: *Geophysics* v. 36, pp. 311-329, 533-553, 676-689.

Maureau, G.T.F.R., and van Wijhe, D.H. (1979), The prediction of porosity in the Permian Zechstein 2 carbonate of eastern Netherlands using seismic data: *Geophysics* v. 44, p. 1502-1517.

Meckel, L.D., and Nath, A.K. (1977), Geologic considerations for stratigraphic modeling and interpretation: *AAPG Memoir 26,* p. 417-438.

Mitchum, R.M. (1977), Glossary of terms used in seismic stratigraphy: *AAPG Memoir 26,* p. 205-212.

Mitchum, R.M., and Vail, P.R. (1977), Seismic stratigraphy interpretation procedure: *AAPG Memoir 26,* p. 135-144.

Mitchum, R.M., Vail, P.R., and Sangree, J.B. (1977), Stratigraphic interpretation of seismic reflection patterns in depositional sequences: *AAPG Memoir 26,* p. 117-134.

Mitchum, R.M., Vail, P.R., and Thompson, S. (1977), The depositional sequence as a basic unit for stratigraphic analysis: *AAPG Memoir 26,* p. 53-62.

Neidell, N.S., and Poggiagliolmi, E. (1977), Stratigraphic modeling and interpretation—geophysical principles and techniques: *AAPG Memoir 26,* p. 389-416.

Payton, C.E. (1977), ed., *Seismic stratigraphy—applications to hydrocarbon exploration: AAPG Memoir 26.*

Ramsayer, G.R. (1979), Seismic stratigraphy, a fundamental exploration tool: *Offshore Technology Conference* paper 3568.

Roksandic, M.M. (1978), Seismic facies analysis concepts: *Geophysical Prospecting* v. 26, p. 383-398.

Sangree, J.B., and Widmier, J.M. (1977), Seismic interpretation of clastic depositional facies: *AAPG Memoir 26,* p. 165-184.

Sangree, J.B., and Widmier, J.M. (1979), Interpretation of depositional facies from seismic data: *Geophysics* v. 44, p. 131-160.

Schramm, M.W., Dedman, E.V., and Lindsey, J.P. (1977), Practical stratigraphic modeling and interpretation: *AAPG Memoir 26,* p. 477-502.

Sheriff, R.E. (1974), Seismic detection of hydrocarbons—the underlying physical principles: *Offshore Technology Conference* paper 2001.

Sheriff, R.E. (1975), Factors affecting seismic amplitudes: *Geophysical Prospecting* v. 23, p.125-138.

Sheriff, R.E. (1976), Inferring stratigraphy from seismic data: *AAPG Bulletin* v. 60, p. 528-542; also in L.W. and D.O. LeRoy, *Subsurface Geology: Petroleum, Mining, Construction,* Colorado School of Mines, Golden, Colorado, p. 438-446.

Sheriff, R.E. (1977a), Limitations on resolution of seismic reflections and geologic detail derivable from them: *AAPG Memoir 26,* p. 3-14.

Sheriff, R.E. (1977b), Using seismic data to deduce rock properties: p. 243-274 in G.D. Hobson, ed., *Developments in Petroleum Geology-I,* Applied Science Publishers, London.

Sheriff, R.E. (1978), *A First Course in Geophysical Exploration and Interpretation:* International Human Resources Development Corporation, Boston.

Sheriff, R.E. (1980), Nomogram for Fresnel-zone calculation: *Geophysics* v. 45, p. 968-992.

Sheriff, R.E., Taner, M.T., and Rao, K. (1978), Seismic attribute measurements in offshore production: *Offshore Technology Conference* paper 3174.

Sheriff, R.E., and Taner, M.T. (1979), Seismic attribute measurements help define reservoirs: *Oil and Gas Journal,* Aug. 20, 1979, p. 182-186.

Shipley, T.H., Buffler, R.T., and Watkins, J.S. (1978), Seismic stratigraphy and geologic history of Blake Plateau and adjacent Western Atlantic continental margin: *AAPG Bulletin* v. 62, p. 792-812.

Sieck, H.C., and Self, G.W., Analysis of high-resolution seismic data: *AAPG Memoir 26,* p. 353-386.

Stommel, H.E., and Graul, M. (1978), Current trends in geophysics, *Indonesian Petroleum Association Proceedings,* Jakarta, Indonesia.

Stone, C.B. (1977), "Bright-spot" techniques: p. 275-291 in G.D. Hobson, ed., *Developments in Petroleum Geology-I,* Applied Science Publishers, London.

Stuart, C.J., and Caughey, C.A. (1977), Seismic facies and sedimentology of terrigenous Pleistocene deposits in northwest and central Gulf of Mexico: *AAPG Memoir 26,* p. 249-276.

Taner, M.T., and Sheriff, R.E. (1977), Application of amplitude, frequency and other attributes to stratigraphic and hydrocarbon determination: *AAPG Memoir 26,* p. 301-328.

Taner, M.T., Koehler, F., and Sheriff, R.E. (1979), Complex seismic trace analysis: *Geophysics* v. 44, p. 1041-1063.

Telford, W.M., Geldart, L.P., Sheriff, R.E., and Keys, D.A. (1976), *Applied Geophysics,* Cambridge University Press, Cambridge, England.

Todd, R.G., and Mitchum, R.M. (1977), Identification of Upper Triassic, Jurassic and Lower Cretaceous seismic sequences in Gulf of Mexico and offshore West Africa: *AAPG Memoir 26,* p. 145-164.

Vail, P.R., and Mitchum, R.M. (1977), Overview of seismic stratigraphy and global changes of sealevel: *AAPG Memoir 26,* p. 49-52.

Vail, P.R., Mitchum, R.M., and Thompson, S. (1977a), Relative changes of sealevel from coastal onlap: *AAPG Memoir 26,* p. 63-82.

Vail, P.R., Mitchum, R.M., and Thompson, S. (1977b), Global cycles of relative changes of sealevel: *AAPG Memoir 26,* p. 83-98.

Vail, P.R., Todd, R.G., and Sangree, J.B. (1977), Chronostratigraphic significance of seismic reflections: *AAPG Memoir 26,* p. 99-116.

Wiemer, R.J., and Davis, T.L. (1977), Stratigraphic and seismic evidence for Late Cretaceous growth faulting, Denver Basin, Colorado: *AAPG Memoir 26,* p. 277-300.

Wyllie, M.R.J., Gardner, G.H.F., and Gregory, A.R. (1962), Studies of clastic wave attenuation in porous media: *Geophysics* v. 27, p. 569-589.

Zieglar, D.L., and Spotts, J.H. (1978), Reservoir and source-bed history of the Great Valley of California: *AAPG Bulletin* v. 62, p. 813-826.

Index